# Other Titles in This Series

44 **J. Amorós, M. Burger, K. Corlette, D. Kotschick, and D. Toledo,** Fundamental groups of compact Kähler manifolds, 1996

43 **James E. Humphreys,** Conjugacy classes in semisimple algebraic groups, 1995

42 **Ralph Freese, Jaroslav Ježek, and J. B. Nation,** Free lattices, 1995

41 **Hal L. Smith,** Monotone dynamical systems: an introduction to the theory of competitive and cooperative systems, 1995

40.2 **Daniel Gorenstein, Richard Lyons, and Ronald Solomon,** The classification of the finite simple groups, number 2, 1995

40.1 **Daniel Gorenstein, Richard Lyons, and Ronald Solomon,** The classification of the finite simple groups, number 1, 1994

39 **Sigurdur Helgason,** Geometric analysis on symmetric spaces, 1993

38 **Guy David and Stephen Semmes,** Analysis of and on uniformly rectifiable sets, 1993

37 **Leonard Lewin, Editor,** Structural properties of polylogarithms, 1991

36 **John B. Conway,** The theory of subnormal operators, 1991

35 **Shreeram S. Abhyankar,** Algebraic geometry for scientists and engineers, 1990

34 **Victor Isakov,** Inverse source problems, 1990

33 **Vladimir G. Berkovich,** Spectral theory and analytic geometry over non-Archimedean fields, 1990

32 **Howard Jacobowitz,** An introduction to CR structures, 1990

31 **Paul J. Sally, Jr. and David A. Vogan, Jr., Editors,** Representation theory and harmonic analysis on semisimple Lie groups, 1989

30 **Thomas W. Cusick and Mary E. Flahive,** The Markoff and Lagrange spectra, 1989

29 **Alan L. T. Paterson,** Amenability, 1988

28 **Richard Beals, Percy Deift, and Carlos Tomei,** Direct and inverse scattering on the line, 1988

27 **Nathan J. Fine,** Basic hypergeometric series and applications, 1988

26 **Hari Bercovici,** Operator theory and arithmetic in $H^\infty$, 1988

25 **Jack K. Hale,** Asymptotic behavior of dissipative systems, 1988

24 **Lance W. Small, Editor,** Noetherian rings and their applications, 1987

23 **E. H. Rothe,** Introduction to various aspects of degree theory in Banach spaces, 1986

22 **Michael E. Taylor,** Noncommutative harmonic analysis, 1986

21 **Albert Baernstein, David Drasin, Peter Duren, and Albert Marden, Editors,** The Bieberbach conjecture: Proceedings of the symposium on the occasion of the proof, 1986

20 **Kenneth R. Goodearl,** Partially ordered abelian groups with interpolation, 1986

19 **Gregory V. Chudnovsky,** Contributions to the theory of transcendental numbers, 1984

18 **Frank B. Knight,** Essentials of Brownian motion and diffusion, 1981

17 **Le Baron O. Ferguson,** Approximation by polynomials with integral coefficients, 1980

16 **O. Timothy O'Meara,** Symplectic groups, 1978

15 **J. Diestel and J. J. Uhl, Jr.,** Vector measures, 1977

14 **V. Guillemin and S. Sternberg,** Geometric asymptotics, 1977

13 **C. Pearcy, Editor,** Topics in operator theory, 1974

12 **J. R. Isbell,** Uniform spaces, 1964

11 **J. Cronin,** Fixed points and topological degree in nonlinear analysis, 1964

(See the AMS catalog for earlier titles)

# Fundamental Groups
of
Compact Kähler Manifolds

# Mathematical
Surveys and Monographs

Volume 44

# Fundamental Groups of Compact Kähler Manifolds

J. Amorós
M. Burger
K. Corlette
D. Kotschick
D. Toledo

**American Mathematical Society**
Providence, Rhode Island

### Editorial Board
Georgia Benkart, Chair     Howard Masur
Robert Greene     Tudor Ratiu

Financial support from the Swiss National Fund, the United States National Science Foundation, the Generalitat de Catalunya, and the Spanish Direccion General de Ciencia Y Tecnologia (grant number PB93-0790) is gratefully acknowledged.

1991 *Mathematics Subject Classification.* Primary 14F35, 32C17, 32J27, 53C55.

ABSTRACT. We study the fundamental groups of compact Kähler manifolds. It is shown that these groups satisfy very strong restrictions arising from Hodge theory, and from its combination with rational homotopy theory, with $L^2$-cohomology, with the theory of harmonic maps and with gauge theory. We also give many examples, some of them new, of groups which arise as fundamental groups of compact Kähler manifolds, in fact, of smooth complex projective varieties; and we consider what happens when one relaxes the Kählerianity assumption on a manifold, or the smoothness assumption on a projective variety.

---

Fundamental groups of compact Kähler manifolds / J. Amorós... [et al.].
    p. cm. — (Mathematical surveys and monographs, ISSN 0076-5376; v. 44)
    Includes bibliographical references and index.
    ISBN 0-8218-0498-7
    1. Kählerian manifolds. 2. Fundamental groups (Mathematics) I. Amorós, J. (Jaume), 1968– . II. Series: Mathematical surveys and monographs; no. 44.
QA649.F77 1996
515′.73–dc20                                                              96-4669
                                                                                                                                                                                      CIP

---

**Copying and reprinting.** Individual readers of this publication, and nonprofit libraries acting for them, are permitted to make fair use of the material, such as to copy a chapter for use in teaching or research. Permission is granted to quote brief passages from this publication in reviews, provided the customary acknowledgment of the source is given.

Republication, systematic copying, or multiple reproduction of any material in this publication (including abstracts) is permitted only under license from the American Mathematical Society. Requests for such permission should be addressed to the Assistant to the Publisher, American Mathematical Society, P.O. Box 6248, Providence, Rhode Island 02940-6248. Requests can also be made by e-mail to `reprint-permission@ams.org`.

© Copyright 1996 by the American Mathematical Society. All rights reserved.
Printed in the United States of America.
The American Mathematical Society retains all rights
except those granted to the United States Government.
∞ The paper used in this book is acid-free and falls within the guidelines
established to ensure permanence and durability.
♻ Printed on recycled paper.

10 9 8 7 6 5 4 3 2 1     01 00 99 98 97 96

# Contents

Preface ix

Chapter 1. Introduction 1
1. Kähler geometry 1
2. Kähler and non–Kähler groups 5
3. Fundamental groups of compact complex surfaces 10
4. Complex symplectic non–Kähler manifolds 15

Chapter 2. Fibering Kähler manifolds and Kähler groups 21
1. The fibration problem 21
2. The Albanese map and free Abelian representations 22
3. Fibering over Riemann surfaces 24
4. Fibering compact complex surfaces 27

Chapter 3. The de Rham fundamental group 29
1. The de Rham fundamental group and the 1–minimal model 29
2. Formality of compact Kähler manifolds 32
3. Applications to the fundamental group and examples 34
4. The Albanese map and the de Rham fundamental group 40
5. Non–fibered Kähler groups 43
6. Mixed Hodge structures on the de Rham fundamental group 45

Chapter 4. $L^2$–cohomology of Kähler groups 47
1. Introduction 47
2. Simplicial $L^2$–cohomology and ends 48
3. de Rham $L^2$–cohomology 51
4. Fibering Kähler manifolds over $\mathbb{D}^2$ 53
5. Fibering Kähler manifolds over Riemann surfaces 60

Chapter 5. Existence theorems for harmonic maps 65
1. Definitions 65
2. Hartman's uniqueness theorem 66
3. The Eells–Sampson theorem 66
4. Equivariant harmonic maps 67

Chapter 6. Applications of harmonic maps 71
1. Existence of pluriharmonic maps 71
2. First applications 76
3. Period domains 81
4. The factorisation theorem 82

| | | |
|---|---|---|
| 5. | Non–linear groups | 85 |
| 6. | Harmonic maps to trees | 87 |

Chapter 7. Non–Abelian Hodge theory 91
1. Basic concepts 91
2. Yang–Mills equations and the $\mathbb{C}^*$–action on Higgs bundles 96
3. Hyperkähler structures and complete integrability 100
4. Applications 103

Chapter 8. Positive results for infinite groups 109
1. Introduction 109
2. The first construction 111
3. A Lefschetz theorem for smooth open varieties 113
4. The general construction 113
5. Non–residually finite Kähler groups 115

Appendix A. Pro group theory 121
1. Definitions of group completions 121
2. Nilpotent completions 124
3. Comparison of nilpotent completions 126

Appendix B. A glossary of Hodge theory 129

Bibliography 133

Index 139

# Preface

*"On ignore à peu près tout de quels groupes peuvent être groupes fondamentaux de variétés algébriques,... "*  P. Deligne (1987)[1]

This book was written because a lot can now be said about the fundamental groups of algebraic varieties and of compact Kähler manifolds. Over the last few years there has been a lot of progress on trying to understand which groups arise. These developments were the topic of the Swiss Seminar (Borel Seminar) in the Spring of 1995. To a large extent, this book is based on lectures given in the seminar, although it is not, and is not meant to be, a faithful account of the seminar. We try to explain what is currently known about the fundamental groups of compact Kähler manifolds[2] and about some closely related questions. A lot of examples are given to show that this class of groups is large and interesting. However, most of the book is devoted to proving restrictions on these groups arising from the work of Johnson–Rees, Gromov, Carlson–Toledo, Simpson and many others. In many cases, especially when good accounts do not already exist, we give complete detailed proofs; in other cases we prove special results and refer the reader to the literature for the general case. The techniques used are a mixture of topology, differential and algebraic geometry, and complex analysis.

Chapter 1, written mostly by D.K., is an introduction and overview, and it explains the context of the problem to which this book is devoted. The section on fundamental groups of compact complex surfaces contains a few new results which have not appeared elsewhere.

Chapter 2, written mostly by D.K., discusses the general problem of finding a holomorphic map inducing a given representation or homomorphism of the fundamental group of a compact Kähler manifold. In general, such a map does not exist, but there are two notable exceptions. The first one is the quotient homomorphism of the fundamental group to its first homology modulo torsion. The Albanese map is a holomorphic map realising this. The second exception is representations onto surface groups of genus at least two. We prove a theorem of Siu to the effect that if the surface group is of maximal genus for the given manifold, then the representation is induced by a surjective holomorphic map with connected fibers from the given Kähler manifold to a Riemann surface. More generally, the property of admitting some holomorphic map onto a closed hyperbolic Riemann surface is seen to be a property of the fundamental group of a compact Kähler manifold. This allows us to divide Kähler groups into so–called fibered and non fibered groups. These concepts tie in nicely with some of the techniques and results in later chapters. We

---
[1] [40], Page 1
[2] Such groups shall be called *Kähler groups* for short.

give complete proofs based on classical arguments (Albanese map, Castelnuovo–de Franchis Theorem), which are toy versions of more delicate arguments in the same style which are used in later chapters.

Chapter 3, written mostly by J.A., applies the techniques of real homotopy theory to study Kähler groups. Rather than looking at the fundamental group itself, we look here at its real Malcev completion. This approach goes back to the work of Sullivan and of Deligne–Griffiths–Morgan–Sullivan in the 1970s, but there are a number of new results as well.

Chapter 4, written mostly by M.B., applies $L^2$–cohomology to prove restrictions on the fundamental groups of Kähler manifolds, following an idea of Gromov and the elaborations on it by Arapura–Bressler–Ramachandran. We give a careful account of Gromov's theorem showing that Kähler groups cannot split as free products. More generally, we show that Kähler groups have finitely many ends. These results are proved by constructing holomorphic fibrations over curves, generalising the results of Chapter 2.

Chapter 5 gives an outline of some existence theorems for harmonic maps. These are needed for the applications in Chapters 6 and 7. First we outline a proof of the theorem of Eells–Sampson, giving existence of harmonic maps in homotopy classes of maps whose target has a non–positively curved Riemannian metric. Then we explain the generalisation of this result to twisted or equivariant harmonic maps due to Corlette, Donaldson and Labourie.

Chapter 6, written mostly by D.T., applies harmonic maps to the study of Kähler groups. It begins with a proof of the Siu–Sampson Bochner formula, which implies that certain harmonic maps are in fact pluriharmonic. Combining this with the existence theorems of Chapter 5, we have a large supply of pluriharmonic maps from Kähler manifolds to negatively curved manifolds. Following Carlson–Toledo, Siu and Sampson, we prove a general factorisation theorem for such maps, which has a number of geometric corollaries. These include a proof of Siu's theorem that was proved using more classical methods in Chapter 2, and many restrictions on Kähler groups. For example, it is shown that fundamental groups of real hyperbolic manifolds of dimension at least three cannot be fundamental groups of compact Kähler manifolds. The final section of this chapter discusses geometric applications of more general harmonic maps, maps for which the target space is a negatively curved space which need not be a manifold. This more general existence theorem is not covered in Chapter 5, and we refer to the original paper by Gromov–Schoen for it.

Chapter 7, written mostly by K.C., is an introduction to the non–Abelian Hodge theory of Corlette and Simpson. This uses the existence theorems for harmonic maps in Chapter 5. There is a detailed discussion of the Riemann surface case due to Hitchin. This motivates the general case, the details of which are often omitted and replaced by references to the original papers. Some applications to fundamental groups are given following Simpson. At the end we present Reznikov's recent proof of the Bloch conjecture.

Chapter 8, written mostly by D.T., gives a number of very non–obvious examples of groups which occur as Kähler groups, in fact, as fundamental groups of smooth complex projective varieties. This includes non–Abelian nilpotent groups, and some groups which are not residually finite. Some of the examples in this chapter have not appeared elsewhere.

There are two appendices summarising standard material used throughout the book. The first appendix, written by J.A. and D.K., presents some generalities about projective completions of finitely generated groups, and the second one serves as a reference for results in Hodge theory.

The main topic *not* covered in this book is the so-called Shafarevich conjecture[3]. At the end of his book [**114**], Shafarevich raised the question whether the universal covering of a smooth complex algebraic variety has to be holomorphically convex, and one can ask the same question for arbitrary compact Kähler manifolds. Over the last few years, the question of Shafarevich has led to intense activities in algebraic geometry. These are obviously related to the study of Kähler groups, but they tend to have a different flavour from the material presented in this book. We refer the reader to the recent monograph by Kollár [**81**] for an overview of this topic. See also [**13**], [**79**].

The 1995 Borel Seminar was organised by D. Kotschick and M. Burger with the cooperation of N. A'Campo and J. Amorós. It met ten times for a total of twenty-five lectures given by N. A'Campo, J. Amorós, M. Burger, F. Campana, J. Carlson, K. Corlette, D. Kotschick, F. Labourie, S. Maier, D. Toledo, A. Valette and K. Zuo. The lecturers at a preparatory meeting in December 1994 were F. Catanese and D. Kotschick. Financial support was provided by the IIIe Cycle Romand de Mathématiques, the Swiss National Fund and the University of Basle.

We are grateful to all the speakers, and to the other participants, for their valuable contributions to the seminar, and, by extension, to our understanding of the topic of this book.

J.A., M.B., K.C., D.K. and D.T.<br>December 1995

Authors' addresses:

J.A.: DEPARTAMENT DE MATEMÀTICA APLICADA I, ETSEIB, UNIVERSITAT POLITÈCNICA DE CATALUNYA, AV. DIAGONAL 647, 08028 BARCELONA, SPAIN

M.B.: INSTITUT DE MATHÉMATIQUES, UNIVERSITÉ DE LAUSANNE, 1015 LAUSANNE, SWITZERLAND

K.C.: DEPARTMENT OF MATHEMATICS, UNIVERSITY OF CHICAGO, 5734 SOUTH UNIVERSITY AVENUE, CHICAGO, ILLINOIS 60637, U.S.A.

D.K.: MATHEMATISCHES INSTITUT, UNIVERSITÄT BASEL, RHEINSPRUNG 21, 4051 BASEL, SWITZERLAND

D.T.: DEPARTMENT OF MATHEMATICS, UNIVERSITY OF UTAH, SALT LAKE CITY, UTAH 84112, U.S.A.

---

[3]In the seminar this was discussed in lectures of F. Campana and K. Zuo.

CHAPTER 1

# Introduction

### 1. Kähler geometry

**1.1. Historical introduction.** In 1932, Erich Kähler submitted a remarkable paper, which appeared the following year [78]. The German title translates as: "On a remarkable Hermitian metric". In this paper, Kähler sets out to derive invariants of a Hermitian metric

$$g = \sum_{i,k=1}^{n} g_{i\bar{k}} dx_i d\bar{x}_k$$

using the then novel calculus of differential forms. The first thing he does is to write down the fundamental 2-form

$$\omega = i \sum_{i,k=1}^{n} g_{i\bar{k}} dx_i \wedge d\bar{x}_k$$

of the metric which we now call the *Kähler form*. Next, he writes down the exterior derivative $d\omega$—this is the first invariant. He says that the case when

(1) $$d\omega = 0$$

"presents itself as a remarkable exception", and immediately starts to write down some consequences, the first of which is that one can express the metric through a potential $u$, now called the *Kähler potential*, in the following manner:

$$g = \sum_{i,k=1}^{n} \frac{\partial^2 u}{\partial x_i \partial \bar{x}_k} dx_i d\bar{x}_k \ .$$

Kähler then points out that the (local) existence of the potential is equivalent to the closedness of $\omega$. At various points he remarks that these notions have an intrinsic meaning[1].

What Kähler did *not* say, although he could have, and we suspect he was aware of it, is that $d\omega = 0$ is also equivalent to the existence of holomorphic coordinates in which the metric has the form

(2) $$g = \sum_{i,k=1}^{n} (\delta_{i\bar{k}} + [2]) dx_i d\bar{x}_k \ ,$$

meaning that up to and including terms of first order, the metric looks like the flat metric on $\mathbb{C}^n$. This immediately implies the following:

SCHOLIUM 1.1. *An identity involving only the metric and its first derivatives holds on any Kähler manifold if and only if it holds in flat $\mathbb{C}^n$.*

---

[1] See [20] for more on the history, and for a discussion of the importance of Kähler's paper for differential geometry.

As a first application of this argument, one sees that equation (2) implies equation (1) which is trivially true in the flat metric. Conversely, a calculation shows that, given (1), one can find new coordinates in which the metric takes the form (2).

Scholium 1.1 has a provocative consequence, which is one of the leitmotives of this book:

METATHEOREM 1.2. *Kähler manifolds are complex manifolds whose geometry reduces to linear algebra.*

This sounds ridiculous, but it has much more substance than one can possibly imagine from a naïve point of view. In fact, not only the local geometry but the global analytic geometry and the algebraic topology of a Kähler manifold are controlled by linear algebra. We will see in Chapter 3 that a lot of potentially quite subtle information about the homotopy type of a Kähler manifold is actually determined by, and can be computed from, the de Rham algebra of differential forms on the manifold (or its cohomology). This algebra is clearly an object of linear algebra, and it is extremely close in spirit to Kähler's original idea of deriving invariants using differential forms.

Here is another example of the reduction to linear algebra, again through differential forms:

THEOREM 1.3 (Castelnuovo–de Franchis (1905)). *Let $X$ be a compact Kähler manifold, and let $\omega_1, \ldots, \omega_r$ be linearly independent holomorphic 1–forms with $\omega_i \wedge \omega_j = 0$ for all $i, j$. Then there exists a holomorphic map $f\colon X \to C$, where $C$ is a complex curve, such that $\omega_1, \ldots, \omega_r$ are in the image of the pullback $f^*$.*

From this one can deduce the following statement, see [**29**], which is even closer to Metatheorem 1.2:

THEOREM 1.4 (Catanese (1989)). *Let $X$ be a compact Kähler manifold. There exists a compact complex curve $C$ of genus $g \geq 2$ and a surjective holomorphic map $f\colon X \to C$ if and only if there exists a $g$–dimensional maximal isotropic subspace $V \subset H^1(X, \mathbb{C})$, where isotropic means that the image of $\Lambda^2(V)$ in $H^2(X, \mathbb{C})$ is zero.*

Perhaps surprisingly, these are really statements about the fundamental group of $X$, as was proved first by Siu [**122**] and later, independently, by Beauville [**10**].

THEOREM 1.5 (Siu (1987), Beauville (1988)). *Let $X$ be a compact Kähler manifold and $g \geq 2$ an integer. Then $X$ admits a non–constant holomorphic map to some compact Riemann surface of genus $g' \geq g$ having connected fibers if and only if there is a surjective homomorphism $h\colon \pi_1(X) \to \pi_1(C_g)$, with $\pi_1(C_g)$ the fundamental group of a compact Riemann surface of genus $g$.*

PROOF. If the holomorphic map exists, it induces a surjection on $\pi_1$, and the surface group of genus $g' \geq g$ surjects onto the surface group of genus $g$.

Conversely, suppose we are given $h$. Then because $C$ is an Eilenberg–Mac Lane space $K(\pi_1(C), 1)$, there exists a homotopy class of continuous maps $f\colon X \to C$ inducing $h$. But the algebra homomorphism $f^*\colon H^*(C, \mathbb{C}) \to H^*(X, \mathbb{C})$ is injective in degree 1 and maps isotropic subspaces of $H^1(C, \mathbb{C})$ to isotropic subspaces of $H^1(X, \mathbb{C})$. Every isotropic subspace is contained in a maximal one. Given the maximal isotropic subspace, Theorem 1.4 shows that there exists a non–constant holomorphic map from $X$ to a curve $C'$. Because one passes from an arbitrary

CHAPTER 1

# Introduction

## 1. Kähler geometry

**1.1. Historical introduction.** In 1932, Erich Kähler submitted a remarkable paper, which appeared the following year [**78**]. The German title translates as: "On a remarkable Hermitian metric". In this paper, Kähler sets out to derive invariants of a Hermitian metric

$$g = \sum_{i,k=1}^{n} g_{i\overline{k}} dx_i d\overline{x}_k$$

using the then novel calculus of differential forms. The first thing he does is to write down the fundamental 2–form

$$\omega = i \sum_{i,k=1}^{n} g_{i\overline{k}} dx_i \wedge d\overline{x}_k$$

of the metric which we now call the *Kähler form*. Next, he writes down the exterior derivative $d\omega$—this is the first invariant. He says that the case when

$$(1) \qquad d\omega = 0$$

"presents itself as a remarkable exception", and immediately starts to write down some consequences, the first of which is that one can express the metric through a potential $u$, now called the *Kähler potential*, in the following manner:

$$g = \sum_{i,k=1}^{n} \frac{\partial^2 u}{\partial x_i \partial \overline{x}_k} dx_i d\overline{x}_k \ .$$

Kähler then points out that the (local) existence of the potential is equivalent to the closedness of $\omega$. At various points he remarks that these notions have an intrinsic meaning[1].

What Kähler did *not* say, although he could have, and we suspect he was aware of it, is that $d\omega = 0$ is also equivalent to the existence of holomorphic coordinates in which the metric has the form

$$(2) \qquad g = \sum_{i,k=1}^{n} (\delta_{i\overline{k}} + [2]) dx_i d\overline{x}_k \ ,$$

meaning that up to and including terms of first order, the metric looks like the flat metric on $\mathbb{C}^n$. This immediately implies the following:

SCHOLIUM 1.1. *An identity involving only the metric and its first derivatives holds on any Kähler manifold if and only if it holds in flat $\mathbb{C}^n$.*

---

[1] See [**20**] for more on the history, and for a discussion of the importance of Kähler's paper for differential geometry.

1

As a first application of this argument, one sees that equation (2) implies equation (1) which is trivially true in the flat metric. Conversely, a calculation shows that, given (1), one can find new coordinates in which the metric takes the form (2).

Scholium 1.1 has a provocative consequence, which is one of the leitmotives of this book:

METATHEOREM 1.2. *Kähler manifolds are complex manifolds whose geometry reduces to linear algebra.*

This sounds ridiculous, but it has much more substance than one can possibly imagine from a naïve point of view. In fact, not only the local geometry but the global analytic geometry and the algebraic topology of a Kähler manifold are controlled by linear algebra. We will see in Chapter 3 that a lot of potentially quite subtle information about the homotopy type of a Kähler manifold is actually determined by, and can be computed from, the de Rham algebra of differential forms on the manifold (or its cohomology). This algebra is clearly an object of linear algebra, and it is extremely close in spirit to Kähler's original idea of deriving invariants using differential forms.

Here is another example of the reduction to linear algebra, again through differential forms:

THEOREM 1.3 (Castelnuovo–de Franchis (1905)). *Let $X$ be a compact Kähler manifold, and let $\omega_1, \ldots, \omega_r$ be linearly independent holomorphic 1–forms with $\omega_i \wedge \omega_j = 0$ for all $i, j$. Then there exists a holomorphic map $f\colon X \to C$, where $C$ is a complex curve, such that $\omega_1, \ldots, \omega_r$ are in the image of the pullback $f^*$.*

From this one can deduce the following statement, see [**29**], which is even closer to Metatheorem 1.2:

THEOREM 1.4 (Catanese (1989)). *Let $X$ be a compact Kähler manifold. There exists a compact complex curve $C$ of genus $g \geq 2$ and a surjective holomorphic map $f\colon X \to C$ if and only if there exists a $g$–dimensional maximal isotropic subspace $V \subset H^1(X, \mathbb{C})$, where isotropic means that the image of $\Lambda^2(V)$ in $H^2(X, \mathbb{C})$ is zero.*

Perhaps surprisingly, these are really statements about the fundamental group of $X$, as was proved first by Siu [**122**] and later, independently, by Beauville [**10**].

THEOREM 1.5 (Siu (1987), Beauville (1988)). *Let $X$ be a compact Kähler manifold and $g \geq 2$ an integer. Then $X$ admits a non–constant holomorphic map to some compact Riemann surface of genus $g' \geq g$ having connected fibers if and only if there is a surjective homomorphism $h\colon \pi_1(X) \to \pi_1(C_g)$, with $\pi_1(C_g)$ the fundamental group of a compact Riemann surface of genus $g$.*

PROOF. If the holomorphic map exists, it induces a surjection on $\pi_1$, and the surface group of genus $g' \geq g$ surjects onto the surface group of genus $g$.

Conversely, suppose we are given $h$. Then because $C$ is an Eilenberg–Mac Lane space $K(\pi_1(C), 1)$, there exists a homotopy class of continuous maps $f\colon X \to C$ inducing $h$. But the algebra homomorphism $f^*\colon H^*(C, \mathbb{C}) \to H^*(X, \mathbb{C})$ is injective in degree 1 and maps isotropic subspaces of $H^1(C, \mathbb{C})$ to isotropic subspaces of $H^1(X, \mathbb{C})$. Every isotropic subspace is contained in a maximal one. Given the maximal isotropic subspace, Theorem 1.4 shows that there exists a non–constant holomorphic map from $X$ to a curve $C'$. Because one passes from an arbitrary isotropic subspace to a maximal one, the genus of $C'$ may be larger than that of $C$. □

In this proof one sees another leitmotiv of this book, namely the consideration of maps between a Kähler manifold $X$ and an Eilenberg–Mac Lane space. This will appear again and again.

We will prove the above Theorems 1.3, 1.4 and 1.5 in Chapter 2. The proof of the Castelnuovo–de Franchis theorem is of course classical, but it will serve as an introduction to the idea of constructing holomorphic maps from foliations defined by holomorphic 1-forms. More sophisticated variants of that argument will appear in Chapters 4 and 6. Siu's theorem gives rise to an important distinction between two kinds of fundamental groups of compact Kähler manifolds, those which surject onto surface groups, and those which do not.

We hope these examples have convinced the reader that Metatheorem 1.2 does have substance. There are many other manifestations of it, for example the Albanese map, cf. Chapter 2, or the theory of variations of Hodge structure (Chapter 7) and the Torelli theorems.

The attentive reader may have observed that the Theorem of Castelnuovo–de Franchis was proved more than a quarter century before Kähler's paper, yet its statement uses the concept of a Kähler metric! This discrepancy is due to the fact that the Italians proved the result for complex algebraic surfaces, but the proof, suitably phrased, works for any compact Kähler manifold.

In fact, smooth complex projective varieties are Kähler manifolds, because $\mathbb{C}P^n$ is Kähler. This fact provides lots of examples, and it is possible that there are, essentially, no others. See the Open Problem 1.7 in the next subsection.

In his paper, Kähler gives various examples of metrics satisfying $d\omega = 0$. They are all negatively curved in some sense; in particular he considers complex balls and polydisks and their quotients, and he points out connections with the theory of automorphic forms. But he does not use the fact that automorphic forms give projective embeddings of the quotients, and he never bothers to look at $\mathbb{C}P^n$ itself.

**1.2. The modern point of view.** From a modern point of view, the Kähler form is defined without recourse to local coordinates. If $J \in End(TX)$, $J^2 = -\text{Id}$, is the almost complex structure on a manifold $X$ with metric $g$, define

$$(3) \qquad \omega(X,Y) = g(JX,Y) .$$

Considering triples $(g, J, \omega)$ satisfying (3) and such that $J$ is $g$- and $\omega$-isometric, any two entries determine the third one. Moreover, $\omega$ is always non-degenerate because

$$\omega^n = n! dvol_g .$$

The Kähler condition $d\omega = 0$ is an integrability condition which makes $\omega$ into a symplectic form. On the other hand, one wants the almost complex structure to be integrable as well, namely one wants it to be induced from holomorphic charts. By the Newlander–Nirenberg Theorem, this is equivalent to the vanishing of the Nijenhuis tensor.

We can now define a Kähler manifold to be one that has compatible complex and symplectic structures. A Riemannian metric is then given automatically by reading (3) backwards. This is the Kähler metric.

Because the Kähler form of a Kähler metric is closed, it represents a cohomology class $[\omega] \in H^2(X, \mathbb{R})$. Moreover, because the top power of $\omega$ is a non-zero multiple of the volume form, this cohomology class, and all its powers up to the top dimension, are non–zero.

Note that it follows from the definitions that every holomorphic submanifold of a Kähler manifold is again Kähler. Thus, to see that smooth complex projective varieties are Kähler, it suffices to show that $\mathbb{C}P^n$ is. We also see that holomorphic submanifolds of compact Kähler manifolds carry non–zero homology classes, because the integral of some power of $\omega$ is a non–zero multiple of the volume and is also a homological intersection number.

One can also describe the Kähler condition completely in terms of submanifold geometry. In fact, $\omega$ has the property that its restriction to any oriented real 2-dimensional submanifold is less than or equal the volume form, and it is equal to the volume form precisely on the holomorphic curves. This means that $\omega$ is a calibration of the complex structure $J$. Harvey–Lawson [67] have proved that the converse is also true, namely every calibrated complex manifold is Kähler.

We construct $\mathbb{C}P^n$ as $(\mathbb{C}^{n+1} \setminus \{0\})/\mathbb{C}^*$. Its complex structure is uniquely determined by the requirement that the projection $\mathbb{C}^{n+1} \setminus \{0\} \to \mathbb{C}P^n$ be holomorphic. This projection restricts to the Hopf map $S^{2n+1} \to \mathbb{C}P^n$. Requiring the differential of the Hopf map to be isometric on the orthogonal complement of the fiber gives $\mathbb{C}P^n$ a unique Riemannian metric $g$. Now we define the Kähler form on $\mathbb{C}P^n$ by formula (3). One has to check that it is closed. Pulling back via the Hopf map, we have $\pi^*\omega = d\alpha$, where $\alpha$ is a 1–form of constant length whose kernel is the orthogonal complement of the fiber of the Hopf map.

The metric we have described is usually rescaled so that
$$\int_{\mathbb{C}P^1} \omega = 1 \ .$$
This means that the cohomology class $[\omega]$ is integral and is a generator of $H^2(\mathbb{C}P^n, \mathbb{Z})$. The rescaled metric is called the *Fubini–Study metric*.

We have seen that smooth complex projective varieties are Kähler manifolds, but the converse is not true. One can have Kähler metrics whose Kähler forms have irrational periods, and so cannot be pull–backs of the Fubini–Study metric. When $\dim H^2(X, \mathbb{R}) > 1$, this problem cannot usually be solved by rescaling. In fact, the integrality of the Kähler class characterises projective algebraic manifolds among all Kähler manifolds:

THEOREM 1.6 (Kodaira Embedding Theorem). *Every compact Kähler manifold whose Kähler form has integral periods can be holomorphically embedded in some $\mathbb{C}P^n$.*

Given such an embedding, a theorem of Chow says that $X$ is an algebraic variety. In complex dimension one, $H^2(X, \mathbb{R})$ is always 1–dimensional, so that all curves are projective. In complex dimension two, Kodaira's classification of compact complex surfaces implies that on every Kähler surface one can deform the complex structure so that, for the deformed structure, $X$ is projective algebraic. In higher dimensions this is not known to be true. On the other hand, nobody knows an example of a Kähler manifold which cannot be deformed to a projective one. This has led many authors to propose that the following problem should have a positive answer:

OPEN PROBLEM 1.7. *Does every compact Kähler manifold deform to one on which the Kähler form has rational periods?*

By rescaling and the Kodaira–Chow package, this would be enough to make such a manifold projective algebraic.

Note that if one drops the integrability condition on the complex structure, one can deform every symplectic form to one with rational periods.

As complex projective algebraic varieties are the main source of examples of Kähler manifolds, it is worthwhile to recall one of the main theorems about their topology:

THEOREM 1.8 (Lefschetz Hyperplane Theorem). *Let $X \subset \mathbb{C}P^n$ be a smooth projective variety and $Y = X \cap H$ a generic hyperplane section. Then for all $i \leq \dim_{\mathbb{C}}(X) - 2$ we have*
$$\pi_i(Y) \cong \pi_i(X) .$$

This shows that in the study of fundamental groups of projective algebraic varieties we can restrict ourselves to complex dimension 2, i.e., to algebraic surfaces. In the Kähler situation there is no such reduction. However, as noted above, the Kodaira classification of compact complex surfaces implies that every Kähler surface deforms to an algebraic one. If the deformation problem has a positive solution in higher dimensions, then the class of fundamental groups of compact Kähler manifolds is the same as the class of fundamental groups of algebraic surfaces.

It should be clear from this discussion that the topology of compact Kähler manifolds is rather restricted, and very interesting to study. Instead of investigating all the topological invariants, we shall concentrate on the fundamental group. This is in some sense the simplest, but also the most interesting homotopy–theoretic invariant. The study of the fundamental group turns out to be very rich, involving topology, algebraic and differential geometry and complex analysis in various guises and combinations.

## 2. Kähler and non–Kähler groups

It has been known[2] at least since the early 1930s that every finitely presentable group is the fundamental group of a closed smooth orientable $n$–dimensional manifold $M$, for all $n \geq 4$. This is not true in dimensions $n < 4$, although every finitely presentable group is the fundamental group of a 2–complex. One can think of the restriction to closed manifolds as a geometric condition which results in constraints on the fundamental group. In all dimensions, imposing some geometric structure on the manifold $M$ often restricts the possible fundamental groups.

In the later chapters of this book we study the case of compact Kähler manifolds, that is, manifolds with compatible complex and symplectic structures. If one relaxes the Kähler condition, it is known that every finitely presentable group is the fundamental group of a closed almost complex, and even symplectic, manifold of real dimension four, and of a complex manifold of real dimension six, which can be taken to be symplectic as well, although the complex and symplectic structures will, of course, not be compatible. We shall discuss these results in Section 4. In Section 3 we discuss the intermediate case of compact complex surfaces, where the fundamental group turns out to be quite restricted, unlike in the non–integrable and in the higher–dimensional cases.

The compactness assumption is essential for the whole discussion. It is easy to build open (Kähler) complex surfaces with preassigned fundamental groups by

---

[2]See the exercise on page 180 of Seifert–Threlfall [**110**] and Section 4.1 below.

thickening 2–complexes. Further, it follows from the work of Eliashberg [**46**] (cf. also [**91**]) that every finitely presentable group is the fundamental group of a Stein surface[3], although not necessarily of an affine algebraic variety, for which real homotopy theory gives restrictions. (See Chapter 3 and [**94**].)

The basic objects we study in this book are Kähler groups.

DEFINITION 1.9. A group is called a *Kähler group* if it is the fundamental group of some compact Kähler manifold.

A finitely presentable group which is not a Kähler group will sometimes be called a *non–Kähler group*.

The next two subsections give examples of both Kähler and non–Kähler groups. The positive examples in fact exhibit certain groups as fundamental groups of smooth complex projective varieties, whereas the negative results give restrictions only using the Kähler condition[4].

**2.1. Positive results.**

EXAMPLE 1.10. The class of Kähler groups is closed under direct products and under passage to finite index subgroups; because a product of Kähler manifolds is Kähler with respect to the product metric, respectively because Kähler metrics can be lifted to coverings[5].

EXAMPLE 1.11. Serre [**112**] proved that every finite group is Kähler. We summarise an argument given in the book [**114**] by Shafarevich.

Note that because we can pass to subgroups using Example 1.10, it is enough to prove that the symmetric groups are Kähler.

Let $S_m$ be the symmetric group on $m$ letters. It acts on the product $\Pi = \mathbb{C}P^s \times \ldots \times \mathbb{C}P^s$ ($m$ factors) by permuting the factors. The quotient $\Pi' = \Pi/S_m$ is a projective variety which is singular only along the image $\Delta' \subset \Pi'$ of the diagonals

$$\Delta = \{(x_1,\ldots,x_m) | x_i = x_j \text{ for some } i \neq j\} \subset \Pi \,.$$

Choose a projective embedding $\Pi' \subset \mathbb{C}P^N$ and intersect $\Pi'$ with a linear subspace $L \subset \mathbb{C}P^N$. If $d = \text{codim}_{\mathbb{C}P^N}(L) > \text{codim}_{\Pi'}(\Delta') = s$, we can choose $L$ so that it does not meet $\Delta'$, and so that $X = L \cap \Pi'$ is smooth. Its preimage in $Y \subset \Pi$ does not meet $\Delta$ and can be taken to be the intersection of $\Pi$ with a linear subspace under a projective embedding of $\Pi$. By the Lefschetz Hyperplane Theorem 1.8, $Y$ is simply connected as long as $\dim(Y) = \dim(\Pi) - \text{codim}(L) = ms - d \geq 2$. The covering group of $Y \to X$ is $S_m$, which shows that $\pi_1(X) \cong S_m$, as desired.

Clearly we can choose $s$ and $d$ so that all the conditions are satisfied.

EXAMPLE 1.12. For all $n$, the integer lattice $\mathbb{Z}^{2n}$ is Kähler, as it is the fundamental group of the Abelian variety $\mathbb{C}^n/\mathbb{Z}^{2n}$.

EXAMPLE 1.13. The fundamental groups of closed orientable 2–manifolds are Kähler, because orientable surfaces have complex structures and in real dimension 2 the Kähler condition $d\omega = 0$ is vacuous.

---

[3]Compare Problem J in [**136**].

[4]Long after this Section was written, we found out that it is very similar to the treatment of Arapura [**4**]. Rather than trying to make it different, we made it more similar by incorporating some remarks from [**4**] that we had missed.

[5]Compare Example 1.19.

EXAMPLE 1.14. Cocompact lattices in the isometry groups of Hermitian symmetric spaces of non–compact type are Kähler. This is clear for torsion–free lattices. For the general case we use the following observation closely related to a remark of Kollár [**4**]:

LEMMA 1.15. *Let $\Gamma$ be a properly discontinuous group of holomorphic automorphisms of a simply connected complex manifold $X$, and assume that there is a subgroup of finite index $\Delta \subset \Gamma$ acting freely and cocompactly preserving a Kähler metric on $X$. Then $\Gamma$ is a Kähler group.*

PROOF. After passing to a subgroup of finite index if necessary, we may assume that $\Delta$ is normal in $\Gamma$. Let $S$ be a compact Kähler manifold with fundamental group $\Gamma/\Delta$, as given by Example 1.11. Then the diagonal action of $\Gamma$ on $X \times \tilde{S}$ is free, and the quotient is a compact complex manifold $Y$ with fundamental group $\Gamma$. Clearly the diagonal $\Gamma$–action preserves a Kähler metric, and so $Y$ is Kähler. □

EXAMPLE 1.16. There exist Kähler groups which are not residually finite. The original examples due to Toledo [**133**], and others, including some new ones, will be explained in Chapter 8.

This last example shows that one should be careful not to make too many assumptions about linear representations when studying Kähler groups, because, by a result of Malcev, non–residually finite groups have no faithful linear representations.

The known examples of Kähler groups which are not residually finite have large pro–finite completions. It would be interesting to know whether there are non–trivial Kähler groups with trivial pro–finite completions. This is a variant of the following:

OPEN PROBLEM 1.17 (Serre). *Is there an irreducible complex algebraic variety with a non–trivial fundamental group with trivial pro–finite completion?*

A concrete instance of this problem, for the Higman 4–group $H$, is raised in [**113**]. This group is given by the following presentation:

(4)
$$H = \langle x_1, x_2, x_3, x_4 \mid x_2 x_1 x_2^{-1} = x_1^2, x_3 x_2 x_3^{-1} = x_2^2, x_4 x_3 x_4^{-1} = x_3^2, x_1 x_4 x_1^{-1} = x_4^2 \rangle .$$

This is an infinite group which has no subgroups of finite index. It follows from results of Gromov–Schoen [**62**], see Chapter 6, that the Higman group is not Kähler. But we do not know whether it is the fundamental group of a singular irreducible variety.

## 2.2. Negative results.

EXAMPLE 1.18. The first Betti number, or rank of the Abelianisation, of a Kähler group has to be even, because for every compact Kähler manifold Hodge theory, cf. Appendix B, implies $b_1 = h^{1,0} + h^{0,1}$ and $h^{1,0} = h^{0,1}$.

EXAMPLE 1.19. Combining Example 1.18 above and Example 1.10, we see that a group $\Gamma$ with a subgroup $\Gamma' \subset \Gamma$ of finite index such that $b_1(\Gamma')$ is odd cannot be Kähler.

This implies that non–trivial free groups cannot be Kähler, because they always have subgroups of finite index and odd rank. It also implies that $\mathbb{Z}_2 * \mathbb{Z}_2$ is not Kähler, because it contains $\mathbb{Z}$ with index 2.

A group may contain a Kähler group as a subgroup of finite index without being itself Kähler. For example, the fundamental group of the Klein bottle (with $b_1 = 1$) cannot be Kähler, although it contains $\mathbb{Z}^2$ with index 2.

EXAMPLE 1.20. Suppose $X$ is a compact manifold, or, more generally, a finite CW–complex. One can build an Eilenberg–Mac Lane space $K(\pi_1(X), 1)$ from $X$ by successively attaching cells of dimensions $3, 4, \ldots$. One has a tautological continuous map $c\colon X \to K(\pi_1(X), 1)$, which, by construction, induces an isomorphism of fundamental groups and has the following property: the algebra homomorphism

$$c^*\colon H^*(\pi_1(X)) \longrightarrow H^*(X)$$

is an isomorphism in degrees $\leq 1$ and is injective in degree 2. It follows that if $\Gamma$ is the fundamental group of a compact Kähler manifold, the pairing coming from the Hard Lefschetz Theorem B.5 recalled in Appendix B is a non–degenerate bilinear pairing on the group cohomology of $\Gamma$, which factors through the cup product:

$$H^1(\Gamma, \mathbb{C}) \times H^1(\Gamma, \mathbb{C}) \longrightarrow H^2(\Gamma, \mathbb{C}) \longrightarrow \mathbb{C}.$$

This shows, for example, that a group $\Gamma$ with $b_1(\Gamma) \neq 0$ but $b_2(\Gamma) = 0$ cannot be Kähler. Further, if $\Delta$ is any group, and $\Gamma$ is as before, then $\Gamma * \Delta$ cannot be Kähler.

EXAMPLE 1.21. Generalising the previous examples, we will see in Chapter 4 that no non–trivial free product is Kähler. This was first proved by Johnson–Rees [75] for free products in which each factor has a non–trivial finite quotient, using the pairing constructed in Example 1.20. The more general result is due to Gromov [58], and has been elaborated on by Arapura–Bressler–Ramachandran [5], who showed that a Kähler group has at most one end, giving an affirmative answer to a question raised by Johnson–Rees [75]. In Chapter 4 we also prove a stronger result from [5] showing that an extension of a group with infinitely many ends by a finitely generated group cannot be Kähler. As an application of this, one sees that the pure braid group on $n$ strands, $P_n$, is not Kähler for any $n$, although its first Betti number $\frac{1}{2}n(n-1)$ is even[6] for half the values of $n$. See Chapter 4, or [4].

It is worth remarking that Gromov's argument using $L^2$–cohomology and the theory of Gromov–Schoen [62] discussed in Chapter 6 are the only arguments restricting Kähler groups which do not make any use of linear representations.

EXAMPLE 1.22. In Chapter 3 we will see that real homotopy theory in the sense of Sullivan strongly restricts Kähler groups. For example, all the Massey triple products of classes in $H^1(\Gamma)$ must vanish. This can be applied in many concrete examples, such as the Heisenberg groups, which are the fundamental groups of Iwasawa manifolds.

EXAMPLE 1.23. In Chapter 6 we will see that the fundamental groups of closed real hyperbolic manifolds of dimension $\geq 3$ cannot be Kähler. The argument, due to Carlson–Toledo [25], is a generalisation of Siu's argument in the proof of Theorem 1.5.

EXAMPLE 1.24. $SL(n, \mathbb{Z})$ is not Kähler, for all $n \geq 2$.

---

[6]The usual braid group $B_n$, which contains the pure braid group as a subgroup of finite index, is not Kähler for any $n$ because its first Betti number is $= 1$.

## 2. KÄHLER AND NON–KÄHLER GROUPS

For $n = 2$ this is easy, because then there are subgroups of finite index which are free, so Example 1.19 applies. Alternatively, we can invoke Example 1.21, because $SL(2, \mathbb{Z})$ has infinitely many ends.

For $n > 2$, this result follows from non–Abelian Hodge theory, see Corollary 7.11 in Chapter 7.

**2.3. The characterisation problem.** In spite of the many interesting results that have been proved about Kähler groups, many of which are discussed in this book, there is no characterisation of these groups, even conjecturally. Note that combining Examples 1.10, 1.11, 1.12 and 1.18, we obtain:

FACT 1.25. A finitely generated Abelian group is Kähler if and only if it has even rank.

In the non–Abelian case no clean statement is in sight; even the case of nilpotent groups is unresolved. See however [27] and [23].

As mentioned earlier, the positive results we have quoted all produce smooth complex projective varieties, whereas the negative results apply to all compact Kähler manifolds. Is is natural to wonder if this is an accident, or not.

OPEN PROBLEM 1.26. *Is every Kähler group the fundamental group of a smooth complex projective variety?*

If the Open Problem 1.7 has a positive solution, then so does this one. We shall see in Chapter 3 that all the real pro–unipotent or torsion–free nilpotent completions of Kähler groups do in fact arise from fundamental groups of smooth projective varieties.

In the other direction, Arapura and Nori [6] recently announced a restriction on the fundamental groups of smooth projective varieties which is not (yet) known to hold for all Kähler groups.

From the point of view of algebraic geometry, one should investigate algebraic varieties which are not necessarily smooth. Given any finite simplicial complex $K$, one can form an affine variety $A$ by replacing each simplex by an affine space of complex dimension equal to the real dimension of the simplex, and a projective variety $P$ by replacing each simplex by a projective space of complex dimension equal to the real dimension of the simplex. In this way one obtains a reducible affine variety $A$ of the same homotopy type as $K$, and a reducible projective variety $P$ with the same fundamental group as $K$. In particular, there are no restrictions on fundamental groups of projective varieties if one admits reducible ones.

Considering irreducible varieties, the situation is less clear–cut. There is at present no known example of a finitely presentable group which is not the fundamental group of an irreducible variety over the complex numbers, if the variety is allowed to be non compact and to have arbitrary singularities.

The situation becomes clearer if one restricts the singularities. Deligne proved in [38] that the rational or real homotopy restrictions on fundamental groups, see Chapter 3, also apply to normal varieties[7]. In a similar vein, M. Ramachandran has pointed out to us that most harmonic theory techniques used to restrict fundamental groups of compact Kähler manifolds work equally well for *normal* projective varieties. The reason is that normality implies that the fibers of a resolution of

---

[7]We owe this reference to D. Arapura.

singularities are connected, and thus a pluriharmonic map of the resolution restricts to a homotopically trivial harmonic map on each fiber, hence is constant on each fiber. Another application of normality then gives a factorisation through the original variety.

### 3. Fundamental groups of compact complex surfaces

The Kodaira classification of smooth compact complex surfaces implies that their fundamental groups are very restricted[8]. It is the purpose of this Section to make these restrictions explicit. It will become clear that in some ways the fundamental groups of non–Kähler complex surfaces are even more restricted than those of Kähler manifolds. From this, we shall derive that some of the constraints on Kähler groups discussed elsewhere in this book also hold for the fundamental groups of all compact complex surfaces.

**3.1. Consequences of the Kodaira classification.** We collect here what we need from Kodaira's classification of compact complex surfaces, and refer the reader to [8] for proofs and for references to the original papers. See also [48].

The first step of the classification of surfaces is by the parity of the first Betti number. The odd case is surprisingly constrained:

THEOREM 1.27 (Kodaira). *Suppose $X$ is a compact complex analytic surface with $b_1(X) \equiv 1 \pmod 2$. Then either $X$ is elliptic, or $b_1(X) = 1$ and the Kodaira dimension of $X$ is negative.*

We will say that a surface is *of class VII* if it has first Betti number $= 1$ and negative Kodaira dimension[9]. The Hopf surfaces, which by definition are finitely covered by $S^3 \times S^1$, are of class $VII$. A number of other examples of class $VII$ surfaces are known, and there are characterisations of some subclasses, but there is no detailed structure theory or a complete classification. See [95] for what is known.

Theorem 1.27, together with Kodaira's results on elliptic surfaces, see subsection 3.3 below, excludes most groups with odd first Betti number $\geq 3$ from being fundamental groups of compact complex surfaces.

The case of even first Betti number is more complex, but we will not need the classification. We will only use the following:

THEOREM 1.28 (Kodaira). *Suppose $X$ is a compact complex analytic surface with $b_1(X) \equiv 0 \pmod 2$. Then $X$ is deformation equivalent to a projective algebraic surface.*

COROLLARY 1.29. *Let $\Gamma$ be a finitely presentable group. The following conditions on $\Gamma$ are equivalent:*
1. *$\Gamma$ is the fundamental group of a smooth complex projective variety.*
2. *$\Gamma$ is the fundamental group of a smooth complex projective surface.*
3. *$\Gamma$ is the fundamental group of a Kähler compact complex surface.*
4. *$\Gamma$ is the fundamental group of a compact complex surface with even first Betti number.*

---

[8]This fact seems to have been overlooked by many authors, see for example Problem J in [136].

[9]This terminology is different from Kodaira's, and from that in [28], but it has become standard and agrees with [8] and [48].

PROOF. (1) $\Rightarrow$ (2) by the Lefschetz hyperplane Theorem 1.8 and because the fundamental groups of projective curves also arise for ruled surfaces over them.

(2) $\Rightarrow$ (3) because smooth projective varieties are Kähler, and (3) $\Rightarrow$ (4) follows from Hodge symmetry.

(4) $\Rightarrow$ (1) follows from Theorem 1.28. $\square$

REMARK 1.30. Combining the above results with later work of Miyaoka, Todorov and Siu shows that every surface with even first Betti number is in fact Kähler.

EXAMPLE 1.31. The free Abelian groups of even rank are fundamental groups of complex projective varieties and therefore of compact complex surfaces. If the rank $r > 4$, taking hyperplane sections of $\frac{r}{2}$-dimensional Abelian varieties produces surfaces with the same fundamental groups. When the rank $r = 2$, the product of an elliptic curve and a line gives a surface.

In view of Theorem 1.27 and Corollary 1.29, restrictions on the fundamental groups of compact complex surfaces can be proved by considering separately the cases when $X$ is Kähler, $X$ is elliptic with $b_1(X) \equiv 1 \pmod{2}$, and $X$ is of class $VII$. We will state some restrictions which are already known in the Kähler case in subsection 3.2 and then discuss the proofs for elliptic surfaces with $b_1(X) \equiv 1 \pmod{2}$ and for surfaces of class VII in subsections 3.3 and 3.4 respectively.

Before proceeding, recall from Example 1.19 that Kähler groups satisfy strong a priori constraints with respect to subgroups of finite index. If $\Gamma$ is the fundamental group of a compact Kähler manifold, then $b_1(\Gamma)$ is even by Hodge symmetry, and the same is true for all subgroups of $\Gamma$ of finite index because the Kähler metric lifts to coverings. For complex surfaces we have:

COROLLARY 1.32. *Let $\Gamma$ be a finitely presentable group with a subgroup $\Gamma_1$ of finite index such that $b_1(\Gamma)$ and $b_1(\Gamma_1)$ have different parities. Then $\Gamma$ cannot be the fundamental group of a compact complex surface.*

PROOF. Suppose $b_1(\Gamma)$ is even. If $X$ is a compact complex surface with $\pi_1(X) \cong \Gamma$, then $X$ is Kähler and, up to deformation, projective. Thus, so is any finite cover, and every finite index subgroup of $\Gamma$ must also have even first Betti number.

Suppose that $b_1(\Gamma)$ is odd. If $X$ is a compact complex surface with $\pi_1(X) \cong \Gamma$, then by assumption $X$ has a finite unramified cover $Y$ with $\pi_1(Y) \cong \Gamma_1$. Thus, $b_1(Y)$ is even and $Y$ is Kähler. But $Y$ has a finite cover $Z$ whose fundamental group $\Gamma_2 \subset \Gamma_1$ is normal in $\Gamma$. As $Y$ is Kähler, so is $Z$. Averaging the Kähler metric on $Z$ under the covering group $\Gamma/\Gamma_2$ of $Z \to X$ we obtain an invariant Kähler metric which descends to $X$. This contradicts the assumption that $b_1(\Gamma)$ is odd. (Compare the proof of Lemma 1.15.) $\square$

EXAMPLE 1.33. Let $F_r$ be the free group on $r > 1$ generators. This contains free normal subgroups of odd rank and so cannot be a Kähler group. If $r$ is even, this also shows that $F_r$ is not the fundamental group of any compact complex surface. When $r$ is odd, all subgroups of finite index are also of odd rank, so the parity argument is insufficient to exclude $F_r$ from being the fundamental group of a complex surface. See example 1.42 for this.

As another application of Corollary 1.32, consider the group $D_\infty = \mathbb{Z}_2 * \mathbb{Z}_2$. This has first Betti number $= 0$ but contains $\mathbb{Z}$ as a subgroup of index 2. Thus it cannot be the fundamental group of any compact Kähler manifold or complex surface. As $D_\infty$ is the only non–trivial free product with finitely many ends, we have:

COROLLARY 1.34. *If the fundamental group of a compact Kähler manifold or complex surface has finitely many ends, it cannot decompose as a non–trivial free product.*

**3.2. Extensions to all compact complex surfaces of results from the Kähler case.** We shall prove the following:

THEOREM 1.35 (Kotschick). *Suppose $\Gamma$ is the fundamental group of a compact complex surface $X$. Then $\Gamma$ cannot be decomposed as a free product $\Gamma_0 * \Gamma_1$ with both factors containing proper subgroups of finite index.*

This was proved for the Kähler case by Johnson–Rees [75]. A more general result, originally proved in [58], [5], will be proved in Chapter 4 of this book, where it is shown that a Kähler group has finitely many ends and (therefore) cannot split as a non–trivial free product, without any assumption on the free factors. We show here that the stronger result holds for all surfaces, except possibly those of class $VII$, for which Theorem 1.35 is the best we can do[10]. However, all known examples of surfaces of class $VII$ have fundamental groups which have finitely many ends and which are freely indecomposable, cf. [95] and subsection 3.4 below.

THEOREM 1.36 (Carlson–Toledo [28]). *Suppose $\Gamma$ is the fundamental group of a compact complex surface $X$. Then $\Gamma$ cannot be the fundamental group of a closed manifold $N$ of constant negative curvature and $\dim(N) > 2$.*

The Kähler case was proved in [25] and will be explained in Chapter 6. We prove here a partial result for the non–Kähler case, but have to refer to [28] for a proof covering all possible cases. However, as with the indecomposability as a free product, the only cases we do not cover are putative surfaces of class $VII$ which are not known to exist, see subsection 3.4. Our arguments, for both Theorems, are exercises in group cohomology, along the lines of [75], [86].

REMARK 1.37. As pointed out in [28], it follows from Theorem 6.24 in Chapter 6 that Theorem 1.36 continues to hold if $N$ is allowed to have variable but (pointwise) strictly $\frac{1}{4}$–pinched negative sectional curvature. This is best possible, as shown by the compact quotients of $\mathbb{C}H^2$.

**3.3. Elliptic surfaces with odd first Betti number.** The following Theorem collects what we need from Kodaira's analysis of elliptic surfaces with odd first Betti number. A modern treatment can be found in [48], Chapter II.

THEOREM 1.38 (Kodaira). *Let $X$ be a minimal elliptic surface with $b_1(X) \equiv 1 \pmod{2}$. Then $X$ has Euler characteristic $\chi(X) = 0$ and its universal cover is diffeomorphic either to $\mathbb{R}^4$ or to $S^3 \times \mathbb{R}$.*

This implies immediately:

COROLLARY 1.39. *If $X$ is an elliptic surface with $b_1(X) \equiv 1 \pmod{2}$, then $\pi_1(X)$ has one or two ends.*

---

[10] It would follow from an extension of the theory of Gromov–Schoen [62], cf. Chapter 6, to the setting of Hermitian harmonic maps à la Jost–Yau that the stronger result is true for all compact complex surfaces.

Using Corollary 1.34, we obtain:

COROLLARY 1.40. *If $X$ is an elliptic surface with $b_1(X) \equiv 1 \pmod 2$, then $\pi_1(X)$ does not split as a free product of two or more non–trivial groups.*

In the case of two ends, $b_1(X) = 1$. Thus, combining Theorems 1.27 and 1.38, we have:

COROLLARY 1.41. *Let $\Gamma$ be a group with $b_1(\Gamma)$ odd and $\geq 3$. If $\Gamma$ is the fundamental group of a compact complex surface, then $\Gamma$ has one end and admits a closed oriented 4–manifold as a $K(\Gamma, 1)$. In particular, it satisfies 4–dimensional oriented Poincaré duality.*

EXAMPLE 1.42. Abelian groups of odd rank $\geq 3$ and free groups of odd rank $\geq 3$ cannot be the fundamental groups of compact complex surfaces.

The next Corollary was also proved by Carlson and Toledo [**28**], who used the characterisation of Abelian subgroups of fundamental groups of negatively curved manifolds.

COROLLARY 1.43. *If $X$ is an elliptic surface with $b_1(X) \equiv 1 \pmod 2$, then $\pi_1(X)$ is not the fundamental group of any closed manifold $N$ with negative sectional curvature.*

PROOF. We may assume that $X$ is minimal. Suppose $N$ is a manifold with negative sectional curvature and $\pi_1(N) \cong \Gamma \cong \pi_1(X)$. Then $\Gamma$ has one end only and both $X$ and $N$ are Eilenberg–Mac Lane spaces for $\Gamma$. Therefore they are homotopy equivalent. Thus $N$ is 4–dimensional and by the Chern–Milnor theorem $\chi(N) > 0$, see [**84**]. This contradicts $\chi(X) = 0$. □

REMARK 1.44. We have given only the crudest information available in Theorem 1.38 because we want to stress that one does not need any detailed analysis for the proofs of the corollaries. In fact, the fundamental groups of elliptic surfaces $X$ with $b_1(X) \equiv 1 \pmod 2$ can be described quite precisely, see for example [**48**], Chapter II.

**3.4. Surfaces of class $VII$.** The known examples of minimal surfaces $X$ of class $VII$ are all Hopf surfaces, or deformations of non–minimal Hopf surfaces, or have $\chi(X) = 0$ and universal covers diffeomorphic to $\mathbb{R}^4$, cf. [**95**]. Thus, their fundamental groups have one or two ends, and the arguments of subsection 3.3 above show that $\pi_1(X)$ is freely indecomposable and is not the fundamental group of any closed manifold of negative curvature.

To prove Theorems 1.35 and 1.36 we have to extend the arguments to cover other, putative, surfaces of class $VII$. For this we need the following constraint on the topology of class $VII$ surfaces:

LEMMA 1.45. *If $X$ is a surface of class $VII$, then so is every finite covering of $X$. Further, the intersection form on $H_2(X, \mathbb{Z})/Tor$ is negative definite, i.e., $b_2^+(X) = 0$.*

PROOF. The Kodaira dimension is invariant under passage to finite coverings, and, by Corollary 1.32, so is the parity of the first Betti number.

As the first Betti number is odd, we have $b_2^+(X) = 2p_g(X)$. Negative Kodaira dimension implies $p_g(X) = 0$. Thus $X$ and all its finite covers have negative definite intersection forms.

Now let $Y$ be a covering of $X$ of degree $d$. Then the multiplicativity of the signature implies $b_2(Y) = db_2(X)$, and the multiplicativity of the Euler characteristic implies $2 - 2b_1(Y) + b_2(Y) = d(2 - 2b_1(X) + b_2(X))$. Combining the two equations and using $b_1(X) = 1$ gives $b_1(Y) = 1$. □

The following Corollary proves Theorem 1.35 for class $VII$ surfaces.

COROLLARY 1.46. *Let $X$ be a surface of class $VII$ and suppose that $\pi_1(X) \cong \Gamma_0 * \Gamma_1$. Then one of the factors, say $\Gamma_0$, has no proper subgroup of finite index, in particular $b_1(\Gamma_0) = 0$. Furthermore, $b_1(\Gamma_1) = 1$ and for every subgroup $\Gamma \subset \Gamma_1$ of finite index $b_1(\Gamma) = 1$ as well.*

PROOF. Suppose that $\pi_1(X) \cong \Gamma_0 * \Gamma_1$ with $b_1(\Gamma_i) = i$. If $\Gamma_0$ has a subgroup $\Gamma$ of finite index $d > 1$, then $\pi_1(X)$ has $\Gamma * (\Gamma_1)^{*d}$ as a subgroup of index $d$. This implies that $X$ has an unramified cover $Y$ of degree $d$ with $b_1(Y) = b_1(\Gamma * (\Gamma_1)^{*d}) \geq d > 1$, contradicting Lemma 1.45. The assertion about subgroups of $\Gamma_1$ follows by the same argument with the rôles of $\Gamma_0$ and $\Gamma_1$ interchanged. □

REMARK 1.47. The Corollary shows that if the fundamental group $\Gamma$ of a surface of class $VII$ decomposes as a free product, then one of the factors, and, a fortiori, $\Gamma$ itself, is not residually finite. A non–residually finite example might look as follows: let $\Gamma = H * \mathbb{Z}$, where $H$ is the Higman four–group of (4), an infinite group with no subgroups of finite index. Then $\Gamma$ and all its subgroups of finite index have first Betti number $= 1$, and $\Gamma$ is the fundamental group of a closed orientable smooth 4–manifold with definite intersection form $X$, in fact, one can take $b_2(X) = 0$. To see this, recall that $H$ has a presentation with 4 generators and 4 relations. Start with the connected sum of 5 copies of $S^1 \times S^3$ and do surgery according to the 4 relations in $H$ to obtain $X$ with $\pi_1(X) \cong \Gamma$. The Euler characteristic of $X$ is $= 0$, and so $b_2(X) = 0$.

We will see in Chapter 8 that there are Kähler groups which are not residually finite, although they are freely indecomposable by the results of Chapter 4.

THEOREM 1.48 ([**28**]). *If $X$ is a surface of class $VII$, then $\pi_1(X)$ is not the fundamental group of any closed hyperbolic manifold $N$.*

The proof by Carlson–Toledo [**28**] uses the theory of Hermitian harmonic maps developed by Jost–Yau. Note however that when $N$ is arithmetic, the result follows immediately from Lemma 1.45, because then one can increase the first Betti number of $N$ by passing to a finite unramified cover by [**16**], section 4.2. (Instead of using the lemma, one could appeal to the fact that the result is already known for complex surfaces with odd first Betti number $> 1$ by Corollary 1.43.) Unfortunately, this does not give a complete proof of Theorem 1.48 because non–arithmetic hyperbolic manifolds $N$ exist (in all dimensions).

When $\dim(N) \leq 4$, one can give very simple proofs of Theorem 1.48. The case $\dim(N) = 2$ is not possible because then the first Betti number is even. The cases $\dim(N) = 3$ or $4$ can be handled using cohomological arguments exploiting properties of the algebra homomorphism induced on cohomology by the classifying map of the universal covering of $X$, as in Example 1.20. Instead of going through all the elementary arguments again, we recall some results from [**86**] which give a clean formulation of what is needed.

DEFINITION 1.49 ([**86**]). For a finitely presentable group $\Gamma$ let
$$p(\Gamma) = inf_X \{\chi(X) - |\sigma(X)|\},$$

where the infimum is taken over all smooth closed oriented 4–manifolds $X$ with $\pi_1(X) = \Gamma$, and where $\chi$ denotes the Euler characteristic and $\sigma$ the signature.

PROPOSITION 1.50 ([86]). *For every finitely presentable group, $p(\Gamma)$ is an even integer satisfying*
$$p(\Gamma) \geq 2 - 2b_1(\Gamma) ,$$
*with equality if and only if $\Gamma$ is the fundamental group of a manifold with (negative) definite intersection form. Furthermore,*
$$p(\Gamma) \leq 2 - 2b_1(\Gamma) + 2a$$
*if and only if $\Gamma$ is the fundamental group of a smooth manifold with $b_2^+ \leq a$.*

Combining this with lemma 1.45 we obtain:

PROPOSITION 1.51. *If $\Gamma$ is the fundamental group of a class VII surface, then $p(\Gamma) = 0$.*

On the other hand, we have the following:

PROPOSITION 1.52 ([86]). *If $\Gamma$ is the fundamental group of a closed orientable aspherical 3–manifold, then $p(\Gamma) = 2$.*

Propositions 1.51 and 1.52 imply Theorem 1.48 in the case $\dim(N) = 3$ because orientability of $N$ can be achieved by passing to a subgroup of index 2 if necessary.

Similarly, the case $\dim(N) = 4$ in Theorem 1.48 follows from the next Proposition.

PROPOSITION 1.53. *If $\Gamma$ is the fundamental group of a closed orientable 4–manifold with $\frac{1}{4}$–pinched negative curvature, then $p(\Gamma) \geq 2$.*

PROOF. Let $N$ be any closed oriented aspherical 4–manifold with fundamental group $\pi_1(N) \cong \Gamma$. Then by theorem 3.8. of [86], we have
$$p(\Gamma) = \chi(N) - |\sigma(N)|.$$
If $N$ has negative sectional curvature, $\chi(N) > 0$ by the Chern–Milnor Theorem [84]. If, in addition, $N$ is $\frac{1}{4}$–pinched, it satisfies the Hitchin–Thorpe inequality $\frac{2}{3}\chi(N) \geq |\sigma(N)|$ by [102], and thus $p(\Gamma) \geq \frac{1}{3}\chi(N) > 0$. □

REMARK 1.54. Carlson–Toledo [28] remark that their proof of Theorem 1.48 remains valid when $N$ is allowed to have variable but (locally) $\frac{1}{4}$–pinched sectional curvature. The proof given here for the case $\dim(N) = 3$ requires no pinching condition whatsoever, in fact $N$ does not need to have negative curvature (as long as it is aspherical). In the case $\dim(N) = 4$ one can relax the pinching assumption in the above Proposition by inspecting the arguments in [102], and in [111], Exposé XI. A lot less is needed to get $\chi(N) > |\sigma(N)|$ so that $p(\Gamma) > 0$.

## 4. Complex symplectic non–Kähler manifolds

We shall show in this Section that every finitely presentable group is the fundamental group of a closed almost–complex, and even symplectic, $2n$–manifold, for every $n \geq 2$, and of a closed complex and symplectic $2n$–manifold, for every $n > 2$. It suffices to prove these results for $n = 2$ and $n = 3$ respectively, because one can then take products with $\mathbb{C}P^1$. These results show that the restrictions on

Kähler groups are not an artefact of the methods of proof via Hodge theory and its variants. Furthermore, the restrictions on fundamental groups of compact complex surfaces discussed in the previous Section represent a low–dimensional phenomenon which has no higher–dimensional analog.

**4.1. Almost complex manifolds.** It is well–known that every finitely presentable group is the fundamental group of a smooth closed orientable $n$–manifold, for every $n \geq 4$. The proof is very easy:

Suppose $\Gamma$ is a group with a presentation with $k$ generators and $l$ relations. The connected sum of $k$ copies of $S^1 \times S^{n-1}$ is a smooth orientable manifold $Y$ with fundamental group the free group on $k$ generators. The relations in the presentation of $\Gamma$ can be represented by $l$ smoothly and disjointly embedded circles in $Y$. Doing elementary surgery on these circles, i.e., replacing tubular neighbourhoods $S^1 \times D^{n-1}$ by $D^2 \times S^{n-2}$, gives a smooth closed orientable manifold $X$. By the Seifert–van Kampen theorem, $\pi_1(X) \cong \Gamma$.

We now improve this to show the following:

LEMMA 1.55 ([85]). *Every finitely presentable group is the fundamental group of a closed almost complex 4–manifold.*

PROOF. By the above discussion, every finitely presentable group is the fundamental group of a closed smooth oriented 4–manifold $X$. We now show how to modify $X$ by connected summing with a simply connected manifold, such that the connected sum is almost complex.

Fix a Riemannian metric $g$ on $X$. If there exists an almost complex structure $J$ on $X$, then we may assume that it is $g$–orthogonal. Then the 2–form $\omega$ defined by equation (3) is a non–vanishing form self–dual under the action of the Hodge star operator defined by $g$. Conversely, if there is a non–vanishing $g$–self–dual form on $X$, the pair $(g, \omega)$ determines an almost complex structure $J$ by equation (3). The existence of $\omega$ is equivalent to the existence of a trivial rank one subbundle of the bundle of self–dual 2–forms:

$$\Lambda_+^2 = \mathbb{R}\omega \oplus K .$$

The orthogonal complement of $\omega$ is an oriented real 2–plane bundle, equivalently a complex line bundle. Its characteristic classes are related to those of $X$ by

$$w_2(X) = w_2(\Lambda_+^2) \equiv c_1(K) \pmod{2} ,$$

$$2e(X) + 3\sigma(X) = p_1(\Lambda_+^2) = c_1^2(K) ,$$

where $\sigma(X)$ denotes the signature of the intersection form of $X$. Thus, the existence of an almost complex structure on $X$ is equivalent to the existence of a class $c = c_1(K) \in H^2(X, \mathbb{Z})$ satisfying the above constraints[11].

Now by replacing $X$ with its connected sum with 4 or 5 copies of $\mathbb{C}P^2$ (with suitably chosen orientations), one can satisfy this criterion, because every positive integer $a \equiv 4 \pmod{8}$ is the sum of the squares of 4 odd integers. (See [85] for the details of this last step.) □

---

[11] This criterion was first found by Wu Wen-Tsun in his Strasbourg thesis.

## 4.2. Symplectic four–manifolds. [12]

The connected sum of two symplectic surfaces is also symplectic. One can exhibit a symplectic structure on the connected sum by explicitly matching some deformations of the original symplectic structures on the annuli used to make the connected sum.

In higher dimensions, the connected sum of two symplectic manifolds is not usually symplectic. In fact, for closed four–manifolds, this never happens. Every symplectic manifold is almost complex, and the connected sum of two closed almost complex four–manifolds is not almost complex. This follows easily from Wu's criterion discussed in the proof of Lemma 1.55. The point is that the second homology is additive under connected sum, and so is the signature, but the Euler characteristic is not.

Gompf [52] discovered that there is a generalisation of the two–dimensional symplectic connected sum construction to higher dimensions, the *symplectic sum* along codimension two symplectic submanifolds. In this construction the connected sum of surfaces is performed fiber-wise in the fibers of the normal bundle of submanifolds. To describe Gompf's construction we need the following lemma, sometimes called *Weinstein's symplectic neighbourhood theorem*.

Suppose $(N^n, \omega_N)$ and $(M^{n+2}, \omega_M)$ are symplectic manifolds, and $\psi \colon N \hookrightarrow M$ is a symplectic embedding with trivial normal bundle. Then:

LEMMA 1.56. *Let $D_r \subset \mathbb{R}^2$ be the disc of radius $r$, with the standard symplectic form $dx_1 \wedge dx_2$. There exist an $\varepsilon > 0$ and a symplectic embedding $\Psi \colon N \times D_\varepsilon \hookrightarrow M$ such that $\Psi|N \times \{0\} = \psi$.*

For $n = 0$ this is just the Darboux theorem. The general case follows from a parametrised version of Moser's technique (which is now a standard argument to prove Darboux's theorem). For details consult [52], Lemma 2.1.

Given symplectic manifolds $(N^n, \omega_N)$, $(M_1^{n+2}, \omega_1)$, $(M_2^{n+2}, \omega_2)$ and symplectic embeddings $\psi_1 \colon N \hookrightarrow M_1$ and $\psi_2 \colon N \hookrightarrow M_2$ with trivial normal bundles, we can find an $\varepsilon > 0$ and $\Psi_1, \Psi_2$ as in the Lemma. Let $\psi \colon D_\varepsilon^* \to D_\varepsilon^*$ be given by $(r, \phi) \mapsto (\sqrt{\varepsilon^2 - r^2}, -\phi)$. Then identify $\Psi_1(N \times D_\varepsilon^*)$ and $\Psi_2(N \times D_\varepsilon^*)$ via $\Psi_1 \circ (Id \times \psi) \circ \Psi_2^{-1}$. The resulting symplectic manifold is called the *symplectic sum* of $M_1$ and $M_2$ along $N$.

To apply this construction in dimension 4 the following observation is useful:

LEMMA 1.57 (Moser). *Let $(N_1, \omega_2)$ and $(N_2, \omega_2)$ be two symplectic surfaces of the same genus and with the same volume. Then every orientation–preserving diffeomorphism $f \colon N_1 \to N_2$ is isotopic to a symplectomorphism.*

PROOF. Note that $[f^*\omega_2 - \omega_1] = 0 \in H^2(N_1, \mathbb{R})$. Thus there is a 1–form $\phi$ with $d\phi = f^*\omega_2 - \omega_1$. If we define $\omega_t = \omega_1 + t(f^*\omega_2 - \omega_1)$, then $\omega_t$ is symplectic for $0 \leq t \leq 1$. Defining the time–dependent vector field $X$ by $X \lrcorner \omega_t = -\phi$, one has

$$L_X \omega_t = d(X \lrcorner \omega_t) + X \lrcorner d\omega_t + \frac{d}{dt}\omega_t = -d\phi + f^*\omega_2 - \omega_1 = 0 \ .$$

Hence, if $\rho_{tt'}$ denotes the flow of $X$ from time $t$ to time $t'$, $\rho_{01}^* f^* \omega_2 = \omega_1$. Thus $f \circ \rho_{01}$ is the desired symplectomorphism. □

COROLLARY 1.58. *Let $\psi_1 \colon N^2 \hookrightarrow M_i^4$ be embeddings with trivial normal bundle into symplectic manifolds $(M_i, \omega_i)$, such that the $\psi^* \omega_i$ are symplectic forms on $N$*

---

[12] This subsection is based on a lecture by S. Maier.

*inducing the same orientation and having the same total volume. Then there is a symplectomorphism* $f : \psi_1(N) \to \psi_2(N)$.

In the situation of the Corollary one can symplectically sum $M_1$ and $M_2$ along $N$. We shall be most interested in the case when $M_2 \setminus \psi_2(N)$ is simply connected. In that case, if $\gamma \subset \psi_1(N)$ is an immersed circle we may perturb it to be disjoint from $\psi_1(N)$. Then $\gamma$ will be contractible in the symplectic sum of $M_1$ and $M_2$ along $N$.

EXAMPLE 1.59. Choose a generic pencil $V$ of cubics in $\mathbb{C}P^2$. The curves of the pencil will intersect in nine points. Blow up these points to obtain the surface $X = \mathbb{C}P^2 \# 9\overline{\mathbb{C}P^2}$ with the projection map $\sigma : X \to \mathbb{C}P^2$. Consider the pencil $\sigma^* V - \sum E_i$, where the $E_i$ are the exceptional curves in $X$. The generic element $N \in \sigma^* V - \sum E_i$ is an elliptic curve with trivial normal bundle, and $X \setminus N$ is simply connected.

We are interested in the following application of the symplectic sum construction:

THEOREM 1.60 (Gompf [52]). *Every finitely presentable group* $\Gamma$ *is the fundamental group of a closed symplectic 4-manifold.*

PROOF. We shall give a complete proof only for the case when $\Gamma \cong F_g$, the free group on $g$ generators. The general case is a slight generalisation of the argument, using the lemma given at the end of the proof.

Let $F$ be a surface of genus $g$. The standard presentation of its fundamental group is obtained by thinking of $F$ as a connected sum of $g$ tori, which we will call $P_1, \ldots, P_g$. Thus

$$\pi_1(F) = \{\alpha_1, \ldots, \alpha_g, \gamma_1, \ldots, \gamma_g \mid \prod_{i=1}^g [\alpha_i, \gamma_i] = 1\},$$

and $\pi_1(F)/<\gamma_1, \ldots, \gamma_g> \cong F_g$.

Collapse $P_1 \# \ldots \# P_{i-1}$ and $P_{i+1} \# \ldots \# P_g$ into $P_i$ and project the resulting torus onto its factor $\gamma_i$. Pulling back the standard 1-form on $S^1$ by this map we get a closed 1-form $\theta_i$ on $F$ which vanishes on each $\gamma_j$, $j \neq i$, and which restricts to a volume form on $\gamma_i$. Let $\theta = \sum_i \theta_i$. Then $\theta$ is closed, and $\theta|\gamma_i$ is a volume form.

Choose symplectic structures on $F$ and on $T^2$, and let $\omega_0$ be the product symplectic structure on $F \times T^2$. Pick any $p \in F \setminus (\bigcup_i \gamma_i)$. Write $T^2 = \alpha \times \beta$ where both $\alpha$ and $\beta$ are standard circles. Let $\theta'$ be a closed 1-form on $T^2$ which restricts to a volume form on $\alpha$. Consider the embedded tori $T_i = \gamma_1 \times \alpha$ and $T_0 = \{p\} \times T^2$. The $T_1, \ldots, T_g$ are Lagrangian submanifolds of $(F \times T^2, \omega_0)$. The form $\theta \wedge \theta'$ is a symplectic form on each of these $T_i$. As nondegeneracy is an open condition, there is an $\varepsilon > 0$ for which $\omega = \omega_0 + \varepsilon(\theta \wedge \theta')$ is symplectic on $F \times T^2$ and restricts to symplectic forms on $T_0, \ldots, T_g$.

For each $i$ let $X_i$ be the elliptic surface of Example 1.59 furnished with the Kähler symplectic form $\omega_i$ scaled such that $\mathrm{vol}(\omega_i|N) = \mathrm{vol}(\omega|T_i)$. Corollary 1.58 allows us to symplectically sum $V_i$ and $F \times T^2$ along $T_i$. The resulting manifold is symplectic with fundamental group $F_g$.

For the general case one uses the following lemma:

LEMMA 1.61 (Gompf [52]). *Let* $\Gamma$ *be a finitely presentable group. There exists a closed oriented 2-manifold* $F$ *with immersed circles* $\gamma_1, \ldots, \gamma_k$ *and a closed 1-form* $\theta$ *on* $F$ *such that:*

1. $\pi_1(F)/<\gamma_1,\ldots,\gamma_k>\cong \Gamma$
2. $\theta|\gamma_i$ is a volume form.

Using this, the proof of the Theorem proceeds as in the special case discussed above. The only difficulty is that the tori $T_i$ may only be immersed but not embedded. A $C^1$–small perturbation of each $\gamma_i$ in $F \times \beta$ will yield disjointly embedded tori, and then one can continue as above. $\square$

REMARK 1.62. Gompf's Theorem 1.60 gives another proof of Lemma 1.55.

**4.3. Complex threefolds.** We will now explain how to deduce from a theorem of Taubes [131] that every finitely presentable group is the fundamental group of a complex manifold of complex dimension 3.

Let $(X^4, g)$ be an oriented Riemannian 4–manifold. Using the Hodge star operator defined by the metric and orientation, we split $\Lambda^2 = \Lambda^2_+ \oplus \Lambda^2_-$. Define the *twistor space* of $(X, g)$ to be $\pi \colon Z \longrightarrow X$, the sphere bundle in $\Lambda^2_+$.

From the discussion in the proof of Lemma 1.55, we know that the fiber of $Z$ over a point $x \in X$ parametrises the orthogonal complex structures on $T_x X$. This means that we can give $Z$ a tautological almost complex structure $J$ as follows: for $z \in Z$, we split $T_z Z = T_z \pi \oplus N_z$, where $T_z \pi$ is the tangent space along the fiber of $\pi$, and $N_z$ is its orthogonal complement with respect to the Levi–Civita connection of $g$. We define $J$ by giving each fiber of $\pi$ its integrable complex structure as the Riemann sphere, and we give each $N_z$, identified with $T_{\pi(z)}X$ by projection, the almost complex structure corresponding to $z$.

Recall that the Riemannian curvature tensor of $(X^4, g)$ decomposes into four components corresponding to the scalar curvature $s$, the trace–less Ricci tensor $Ric_0$, and to the self–dual and anti–self–dual Weyl tensor $W_\pm$ respectively.

The Penrose twistor correspondence gives the following relation between 4–dimensional Riemannian geometry and 3–dimensional complex analysis:

THEOREM 1.63 ([7]). *The almost complex structure $J$ on $Z$ is integrable if and only if $W_+ = 0$.*

EXAMPLE 1.64. An obvious example is $\mathbb{R}^4$ with the flat metric. In this case $Z = \mathbb{R}^4 \times S^2$, with the holomorphic structure of the bundle $\mathcal{O}(1) \oplus \mathcal{O}(1) \longrightarrow \mathbb{C}P^1$.

Another example is $\overline{\mathbb{C}P^2}$ (the complex projective plane endowed with the non–complex orientation) with the Fubini-Study metric. In this case the twistor space $Z$ is the flag manifold $\mathcal{F}_3$ which consists of pairs $(p, x) \in \overline{\mathbb{C}P^2} \times \mathbb{C}^3$ where $x$ is contained in the line defined by $p$.

It is very hard in general to find Riemannian metrics satisfying $W_+ = 0$. However, we have the following deep and difficult result:

THEOREM 1.65 (Taubes [131]). *Let $X$ be any smooth closed oriented 4–manifold. For $n$ sufficiently large $X \# n\overline{\mathbb{C}P^2}$ admits a metric with $W_+ = 0$.*

Unlike Gompf's construction of symplectic manifolds, which is essentially elementary and is of a topological nature, Taubes's construction is part of hard analysis. Both in technique and in difficulty, it is comparable to, but more difficult than, Taubes's existence theorems for instantons in non–Abelian gauge theory.

COROLLARY 1.66. *Every finitely presentable group is the fundamental group of a complex threefold, which is also symplectic.*

PROOF. Let $\Gamma$ be a finitely presentable group. Starting with any smooth closed oriented 4–manifold $Y$ with $\pi_1(Y) \cong \Gamma$, we can find a complex threefold with the same fundamental group as the twistor space $Z$ of $X = Y \# n\overline{\mathbb{C}P^2}$.

It was Gompf [**52**] who first observed that one can take $Z$ to be simultaneously complex and symplectic, although the two structures will of course not usually be compatible. His construction is as follows: we can start with a symplectic $Y$, this is possible by Theorem 1.60. Then $X$ can be taken to be symplectic as well, because one can blow up points symplectically.

By the Thom isomorphism theorem, there is a closed 2–form $\eta_1$ on the twistor space $Z$ such that $\int_{Z_x} \eta = 1$. It is not hard to see that we can take $\eta$ to be everywhere non–zero (compare Lemma 2.2 in [**52**]). Then for all small enough $t > 0$, the form $\pi^*\omega + t\eta$ on $Z$ is symplectic. $\square$

CHAPTER 2

# Fibering Kähler manifolds and Kähler groups

## 1. The fibration problem

If $f\colon X \to Y$ is a surjective holomorphic map between compact complex manifolds, then the image of $\pi_1(X)$ in $\pi_1(Y)$ under the induced map $f_*$ is a subgroup of index bounded by the number of components of the generic fiber of $f$. In particular, if $f$ has connected fibers, $f_*$ is surjective.

We are interested in the converse problem. Suppose $X$ is a compact Kähler manifold with $\pi_1(X) \cong \Gamma$, and $\rho\colon \Gamma \to \Delta$ is a surjective homomorphism, or, more generally, some representation in a finitely presentable group. Then it would be useful to have a surjective holomorphic map $f\colon X \to Y$ with $f_* = \rho$. This would be particularly useful if $Y$ (and $f$) were essentially determined by $\rho$.

As stated, this problem is much too general and cannot be solved. First of all, by the discussion in Chapter 1, the smallest dimension of a complex manifold with fundamental group $\Delta$ is usually 3, whereas $X$ may well be a surface, or even a curve. Thus, even if a holomorphic map exists which induces $\rho$, it will not usually be surjective. Furthermore, there seems to be no natural choice for the target manifold $Y$. Even if we restrict ourselves to the case when $\Delta$ is a Kähler group, there will in general be many candidates for $Y$. For some choices of $\Delta$, $Y$ can be chosen to be an Eilenberg–Mac Lane space, which guarantees at least the existence of a continuous map from $X$ to $Y$ inducing $\rho$. However, this map cannot in general be chosen to be holomorphic, nor does it have a chance of being surjective, simply for dimension reasons.

There is one very important case where our problem does have a positive solution. This is when $\rho\colon \Gamma \to \Delta$ is a surjective homomorphism to a surface group of genus $\geq 2$. In this case there are compact complex curves with fundamental group $\Delta$, and they are aspherical. It will turn out that the existence of a surjective holomorphic map to some curve of genus $\geq 2$ is equivalent to the existence of a surjective homomorphism of $\pi_1(X)$ onto a surface group of genus $\geq 2$. Further, every surjection $\rho$ as above can be induced by a holomorphic map if the genus of the surface group is assumed to be maximal for $\Gamma$, i.e., $\Delta \cong \pi_1(C_g)$ where $C_g$ is a compact Riemann surface of genus $g$, and there is no surjective homomorphism from $\Gamma$ to the fundamental group of any compact Riemann surface of genus $> g$. This will be proved in Section 3 below. As a warm-up, we discuss in the next Section a parallel result for the case where $\Delta$ is free Abelian. This is also a toy case for some of the material in later chapters.

The problem we discuss here has far-reaching generalisations. Given a representation $\rho\colon \pi_1(X) \to G$, one can ask whether there is a holomorphic map $f\colon X \to Y$, where $Y$ is a Kähler manifold of smaller dimension than $X$, such that $\rho$ factors through $f$. We will not address this question here. The interested reader can find some results in this direction in [**138**] and in the papers cited there.

Most of the arguments in this chapter are classical, the exceptions are based on [**122**] and [**29**]. Refinements and generalisations of these arguments will appear throughout the later chapters.

This chapter is based in part on lectures of N. A'Campo and F. Catanese, who contributed greatly to our appreciation of these results.

## 2. The Albanese map and free Abelian representations

Let $X$ be a compact Kähler manifold. The *Albanese variety* of $X$ is the complex torus
$$Alb(X) = H^0(X, \Omega_X^1)^*/j(H_1(X, \mathbb{Z})),$$
where $j$ is the homomorphism
$$j \colon H_1(X, \mathbb{Z}) \longrightarrow H^0(X, \Omega_X^1)^*$$
$$[\gamma] \longmapsto \left(\omega \mapsto \int_\gamma \omega\right).$$

By Hodge theory, cf. Appendix B, the image $\Lambda = \mathrm{Im}(j) \cong H_1(X, \mathbb{Z})/\text{torsion}$ is a lattice of rank $b_1(X) = 2h^0(X, \Omega_X^1)$, and thus $Alb(X)$ is a complex torus. Fixing a basepoint $x_0 \in X$, one defines the *Albanese map* by
$$\alpha_X \colon X \longrightarrow Alb(X)$$
$$x \longmapsto \left(\omega \mapsto \int_{x_0}^x \omega\right).$$

The dependence on the basepoint is very mild; two different choices give maps which are related by a translation in $Alb(X)$. By Hodge theory, $(\alpha_X)_* \colon H_1(X, \mathbb{Z})/\text{torsion} \to H_1(Alb(X), \mathbb{Z})$ is an isomorphism.

The Albanese map is characterised by the following universal property: every holomorphic map from $X$ to a complex torus factors through it in a unique way.

Consider now a surjective homomorphism $\rho \colon \Gamma \to \mathbb{Z}^{2k}$. We want to know whether there is a holomorphic map from $X$ to a complex torus of dimension $k$ inducing $\rho$. The torus is aspherical, so there is always a continuous map to a smooth torus inducing $\rho$. However, the following example shows that this map cannot in general be chosen to be holomorphic, no matter which complex structure one chooses on the torus.

EXAMPLE 2.1. Let $X$ be an Abelian surface not isogenous to a product of elliptic curves. Then $X$ has no non–constant holomorphic maps to elliptic curves, and thus no projection $\pi_1(X) \cong \mathbb{Z}^4 \to \mathbb{Z}^2$ is induced by a surjective holomorphic map from $X$ to a torus.

Thus, our problem has a negative answer in general. However, it has a positive answer if we assume that the first Betti number of $X$ is $2k$, i.e., $k$ is chosen to be maximal for the given fundamental group $\Gamma \cong \pi_1(X)$. This is an immediate consequence of the existence of the Albanese map.

LEMMA 2.2. *Let $X$ be a compact Kähler manifold, and $\rho \colon \pi_1(X) \to \mathbb{Z}^{b_1(X)}$ a surjective homomorphism. Then there is a holomorphic map from $X$ to a complex torus of dimension $q(X) = \frac{1}{2}b_1(X)$ inducing $\rho$.*

PROOF. We may identify $\mathbb{Z}^{b_1(X)} = \pi_1(Alb(X))$. The Albanese map induces a surjection $(\alpha_X)_*\colon \pi_1(X) \to \pi_1(Alb(X))$ which may differ from $\rho$. However, there is always a $\phi \in GL(b_1(X), \mathbb{Z})$ with $\phi \circ (\alpha_X)_* = \rho$, and $\phi$ is realised by a holomorphic map, which we also call $\phi$, so that $\rho$ is induced by $\phi \circ \alpha_X$. □

We will now see how to extract an elementary homotopy invariant of Kähler manifolds from the Albanese map.

DEFINITION 2.3. For any topological space $X$ of finite type (a compact manifold, say, or a finite CW–complex) define $a(X)$ to be the maximal integer $m$ for which the image of $\Lambda^m H^1(X, \mathbb{R})$ in $H^m(X, \mathbb{R})$ is non–trivial.

It is clear that we could use cohomology with rational or complex coefficients to obtain the same invariant as in the real case. The invariant $a(X)$ is bounded above by $\min\{b_1(X), \dim(X)\}$.

By definition, $a(X)$ is a homotopy invariant of $X$, or, more crudely, an invariant of its real cohomology algebra. For Kähler manifolds, the cohomology algebra actually determines the *real* homotopy type, as we will see in Chapter 3.

PROPOSITION 2.4 ([29]). *Let $X$ be a compact Kähler manifold. Then $a(X)$ is the real dimension of its Albanese image $\alpha_X(X)$.*

PROOF. Let $d$ be the real dimension of the Albanese image of $X$.

We use cohomology with complex coefficients to determine $a(X)$. By Hodge theory, cf. Appendix B, $H^1(X, \mathbb{C}) = H^0(\Omega_X^1) \oplus \overline{H^0(\Omega_X^1)}$. Hence $\Lambda^{d+1} H^1(X, \mathbb{C})$ is zero in cohomology if $\Lambda^{\frac{1}{2}d+1} H^0(\Omega_X^1)$ is. The latter is zero in cohomology because it is pulled back from the Albanese variety of $X$ and it involves more holomorphic 1–forms than the complex dimension of the Albanese image.

It remains to show that $\Lambda^d H^1(X, \mathbb{C})$ is non–zero in cohomology. On the Albanese variety we can choose $\frac{1}{2}d$ linearly independent holomorphic 1–forms $\alpha_1, \ldots$, $\alpha_{\frac{1}{2}d}$ such that $\alpha = (\alpha_X)^*(\alpha_1 \wedge \ldots \wedge \alpha_{\frac{1}{2}d})$ is non–zero in $H^0(\Omega_X^{\frac{1}{2}d})$. Then $\alpha \wedge \bar{\alpha}$ represents a non–zero element in the image of $\Lambda^d H^1(X, \mathbb{C})$ in $H^d(X, \mathbb{C})$. □

Though $a(X)$, the Albanese dimension of a compact Kähler manifold $X$, is an invariant of its real homotopy type, it is not an invariant of its fundamental group. Finitely presentable groups do not usually have Eilenberg–Mac Lane spaces of finite type, but we can extend Definition 2.3 to aspherical spaces to obtain a group invariant

$$a(\Gamma) = a(K(\Gamma, 1)) \ .$$

The classifying map of the universal covering of a manifold induces an algebra homomorphism in cohomology which is an isomorphism in degrees $\leq 1$ and injective in degree 2, cf. Example 1.20. This shows that we always have $a(X) \leq a(\pi_1(X))$. We do not usually have equality, because the classifying map is not injective in higher degrees.

EXAMPLE 2.5. For any Abelian variety, the Albanese map is the identity, and $a(X) = \dim_{\mathbb{R}}(X)$. On the other hand, there are algebraic surfaces with fundamental group $\mathbb{Z}^{2n}$ for every $n$, e.g. iterated hyperplane sections of (real) $2n$–dimensional Abelian varieties. For surfaces, $a(X) \leq 4$, whereas $a(\mathbb{Z}^k) = k$.

## 3. Fibering over Riemann surfaces

For every topological space of finite type we can define an integral invariant $g(X)$, called *the genus of $X$*, which, like the $a(X)$ considered in the previous Section, is an invariant of the real cohomology algebra of $X$. However, this genus will turn out to depend only on the fundamental group of $X$. For compact Kähler manifolds, the genus has the useful property of governing the existence of surjective holomorphic maps to Riemann surfaces of genus $\geq 2$.

A subspace $U \subset H^1(X)$ will be called *isotropic* if the natural map $\Lambda^2 U \to H^2(X)$ is identically zero. For compact Kähler manifolds, we have $H^1(X, \mathbb{C}) = H^0(\Omega_X^1) \oplus \overline{H^0(\Omega_X^1)}$, so we can talk of isotropic subspaces in the space of global holomorphic 1–forms.

DEFINITION 2.6. The *genus of $X$* is:
$$g(X) = \max\{\dim U \mid U \subset H^1(X, \mathbb{R}), U \text{ isotropic}\}.$$

The properties of the tautological map
$$c \colon X \longrightarrow K(\pi_1(X), 1)$$
discussed in Example 1.20 show that $g(X) = g(\pi_1(X))$ is really an invariant of the fundamental group. Like $a(X)$, $g(X)$ does not depend on the coefficient field used for cohomology, as long as it has characteristic zero.

To see how the genus controls the existence of surjective holomorphic maps from Kähler manifolds to compact Riemann surfaces, we will need the following classical result:

THEOREM 2.7 (Castelnuovo–de Franchis (1905)). *Let $X$ be a compact Kähler manifold admitting a maximal isotropic subspace $U \subset H^0(X, \Omega_X^1)$ of dimension $g \geq 2$. There exists a surjective holomorphic map $f \colon X \to C$ to a compact Riemann surface $C$ of genus $g$ such that $U$ is in the image of the pullback map $f^*$, and $f$ has connected fibers.*

PROOF. Choose a basis $\omega_1, \ldots, \omega_g \in U$, and let $Y \subset X$ be the non–empty open subset where not all $\omega_i$ vanish. At every point $p \in Y$ we have a distinguished subspace $F_p \subset T_p X$ defined by
$$F_p = \bigcap_{1 \leq i \leq g} \ker(\omega_i(p)).$$
As $U$ is isotropic, we have $\omega_i \wedge \omega_j = 0$, which implies that for all $i, j$, if $\omega_i(p) \neq 0 \neq \omega_j(p)$, then $\ker \omega_i(p) = \ker \omega_j(p)$. Thus $\dim F_p = \dim X - 1$. At each point in $Y$ the distribution $F$ is locally defined by a closed form $\omega_i$ and thus it is integrable. We will see that the leaves of the foliation on $Y$ defined by the distribution $F$ are contained in the level sets of a non–constant holomorphic map from $X$ to a curve.

First, we write $\omega_i = \phi_i \omega_1$ and consider the holomorphic map
$$\phi \colon Y \longrightarrow \mathbb{C}P^{g-1}$$
$$p \longmapsto [1 \colon \phi_2(p) \colon \ldots \colon \phi_g(p)].$$
This is well–defined as a holomorphic map on all of $Y$, although the notation suggests that we need $\omega_1 \neq 0$.

The fact that the $\omega_i$ are closed implies $d\phi_i \wedge \omega_1 = 0$, and so the $\phi_i$ are constant on the leaves of the foliation defined by the $\omega_i$. The image of $\phi$ is thus at most 1–dimensional. The assumption $\dim(U) = g \geq 2$ shows that it is a curve $D$.

Suppose $\phi$ is not holomorphic. Then we pass to a blowup $\pi\colon X' \to X$, such that $\phi$ induces a holomorphic map $\phi'\colon X' \to D$. Let $\psi\colon X' \to C$ be its Stein factorisation. Then $\omega_i = \phi_i\omega_1 = \frac{\phi_i}{\psi_i}d\phi_i$ is a rational 1–form pulled back from some $\eta_i$ on $C$. If $\eta_i$ is not regular on $C$, then $\omega_i$ would have poles, which is a contradiction.

The $\eta_1, \ldots, \eta_g$ are linearly independent in $H^0(\Omega^1_C)$, and thus $g(C) \geq g$.

The components of the exceptional locus of $\pi\colon X' \to X$ are $\mathbb{C}P^1$–bundles, but $\psi$, being a holomorphic map to a hyperbolic Riemann surface, must be constant on each $\mathbb{C}P^1$. Thus $\psi$ (and $\phi'$) descends to define a holomorphic map from $X$ to $C$ (respectively $D$). $\square$

REMARK 2.8. Theorem 2.7 holds, with the same proof, whenever one has an isotropic space of *closed* holomorphic 1–forms on a complex manifold. (In the Kähler case, the closedness is automatic.) On any compact complex surface all holomorphic 1–forms are closed, cf. [**8**].

Instead of blowing up and taking the Stein factorisation, one can prove first that $\phi$ has no indeterminacy. That argument will be presented, and used, in Section 5 of Chapter 4, in a non–compact situation where Stein factorisation is not available. Related arguments are also used in Chapter 6.

COROLLARY 2.9 (Catanese [**29**]). *Let $X$ be a compact Kähler manifold with $g(X) \geq 2$. Then there exists a surjective holomorphic map $f\colon X \to C$ with connected fibers, with $C$ a compact Riemann surface of genus $g(X)$.*

*Moreover, every maximal isotropic subspace $V \subset H^1(X, \mathbb{R})$ is realised as $f^*(V')$ for such a map $f$.*

PROOF. Let $V \subset H^1(X, \mathbb{R})$ be a maximal isotropic subspace. Choose harmonic forms $\varphi_1, \ldots, \varphi_{g(X)}$ representing a basis of $V$.

By Hodge theory we can write $\varphi_i = \omega_i + \bar{\omega}_i$, with $\omega_i \in H^0(X, \Omega^1_X)$. The Hodge type decomposition of the product $[\varphi_i \wedge \varphi_j] = 0 \in H^2(X, \mathbb{C})$ shows that $\omega_i \wedge \omega_j = 0$ for all $i, j$.

Linear independence over $\mathbb{R}$ of the real cohomology classes of $\varphi_1, \ldots, \varphi_{g(X)}$ implies that they are also linearly independent over $\mathbb{C}$ in $H^1(X, \mathbb{C})$. Hence the holomorphic forms $\omega_1, \ldots, \omega_{g(X)}$ form an isotropic subspace $U$ of complex dimension $g(X)$ in $H^0(X, \Omega^1_X)$. By the theorem of Castelnuovo–de Franchis 2.7, from $U$ we can find a surjective holomorphic map $f\colon X \to C$ with connected fibers, and with $g(C) = g(X)$, such that $V$ is in the image of the pullback $f^*$. $\square$

REMARK 2.10. Catanese and others, cf. [**29**] and the papers cited there, have given analogous criteria for the existence of holomorphic maps to higher–dimensional tori. These use the Albanese map and, like the discussion in Section 2 above, depend on properties of the cohomology algebra of $X$ which are not controlled by the fundamental group.

Finally we can prove the theorem of Siu and Beauville showing how the existence of holomorphic fibrations over Riemann surfaces is controlled by the fundamental group.

THEOREM 2.11 (Siu [**122**], Beauville [**10**]). *Let $X$ be a compact Kähler manifold and $g \geq 2$ an integer. Then $X$ admits a surjective holomorphic map to some compact Riemann surface of genus $g' \geq g$ having connected fibers if and only if there* *is a surjective homomorphism* $h \colon \pi_1(X) \to \pi_1(C_g)$, *with* $\pi_1(C_g)$ *the fundamental group of a compact Riemann surface of genus* $g$.

*If* $\rho \colon \pi_1(X) \to \pi_1(C)$ *is a surjective homomorphism, and* $g(C) = g(X) \geq 2$, *then there is a surjective holomorphic map with connected fibers from* $X$ *to some compact Riemann surface of genus* $g(X)$ *inducing* $\rho$.

PROOF. We already saw in Section 1 of Chapter 1 how to deduce this from Corollary 2.9.

If the holomorphic map $f \colon X \to C'$ exists, is non–constant and has connected fibers, then it induces a surjective homomorphism $f_* \colon \pi_1(X) \to \pi_1(C')$. As $g(C') \geq g(C)$, $\pi_1(C')$ surjects onto $\pi_1(C)$.

Conversely, if a surjective homomorphism $\rho \colon \pi_1(X) \to \pi_1(C)$ is given, it can be realised by a continuous map $h \colon X \to C$. Under $h$ isotropic subspaces pull back to isotropic subspaces, and Corollary 2.9 produces a holomorphic map from every maximal isotropic subspace. It is possible that the target of this holomorphic map has genus strictly larger than $g(C)$, if the pullbacks of maximal isotropic subspaces from $C$ are not maximal for $X$.

When $g(C) = g(X)$, the target $C'$ of the holomorphic map given by Corollary 2.9 has genus $g(X)$. If we identify the fundamental groups of $C$ and $C'$ suitably, this holomorphic map induces $\rho$. □

Theorem 2.11 shows that surjective homomorphisms onto maximal surface groups are realised by holomorphic maps, although this may fail without the maximality assumption on $g(C)$. This assumption is analogous to the one made in Lemma 2.2.

We have followed Catanese [**29**] in the presentation of Theorem 2.11. It is worthwhile to explain Siu's original approach in [**122**], because it was the beginning of the applications of harmonic maps to the study of Kähler groups that will be taken up in Chapter 6.

Suppose we are given a surjective homomorphism $\rho \colon \pi_1(X) \to \pi_1(C)$, where $X$ is compact Kähler and $C$ is a compact surface of genus $g(C) \geq 2$. As before, we can choose a homotopy class of continuous maps $h \colon X \to C$ inducing $\rho$. If we choose on $C$ a metric of constant negative curvature, we can apply the Eells–Sampson theorem to find a harmonic representative for $h$, cf. Chapter 5. As $h$ induces a surjection on fundamental groups, its image cannot be a closed geodesic. A factorisation theorem for harmonic maps to hyperbolic space forms, Theorem 6.21 which will be proved in Chapter 6, then shows that $h$ factorises as

$$X \xrightarrow{\varphi} S \xrightarrow{\psi} C,$$

where $S$ is a compact Riemann surface, $\varphi$ is holomorphic and $\psi$ is harmonic. Taking the Stein factorisation of $\varphi$ gives a surjective holomorphic map with connected fibers, whose image is a compact Riemann surface of genus $\geq g(S) \geq g(C)$. If $g(C)$ was maximal to begin with, i.e. $g(C) = g(X)$, then $\varphi$ must have connected fibers and $g(S) = g(C)$. Identifying $\pi_1(S)$ with $\pi_1(C)$ via $\psi_*$, we see that $\rho$ is induced by $\varphi$.

In view of Corollary 2.9 and Theorem 2.11, a compact Kähler manifold admits a surjective holomorphic map to a Riemann surface of genus $\geq 2$ if and only if its fundamental group has a surjective homomorphism to a surface group of genus $\geq 2$, if and only if the genus of $X$ is at least 2. Thus, the following definition is meaningful:

DEFINITION 2.12. Let $\Gamma$ be a Kähler group. We say that it is fibered if $g(\Gamma) \geq 2$, i.e., if every compact Kähler manifold with fundamental group $\Gamma$ fibers over a compact Riemann surface of genus $\geq 2$.

The destinction between fibered and non–fibered Kähler groups will be important in many of the results in later chapters.

REMARK 2.13. If a group $\Gamma$ admits a surjective homomorphism onto a free group $F_r$, then $H^1(F_r, \mathbb{R})$ maps to an isotropic subspace of dimension $r$ in $H^1(\Gamma, \mathbb{R})$, and hence $g(\Gamma) \geq r$. Theorem 2.11 shows that if $\Gamma$ is Kähler and $r \geq 2$, then the homomorphism from $\Gamma$ to $F_r$ can be lifted to the surface group $\pi_1(C_r)$. Thus, there is an equivalent formulation of Definition 2.12, saying that $\Gamma$ is fibered if and only if it surjects onto a free group of rank $\geq 2$.

REMARK 2.14. The examples of Abelian surfaces mentioned in Example 2.1 show that the existence of a surjective holomorphic map to a compact Riemann surface of genus 1 is not invariant under deformations of the complex structure. A fortiori, it is not a property of the cohomology algebra, or of the fundamental group.

The Kähler assumption cannot be dropped from the results in this section. There are many compact complex manifolds whose genus is $\geq 2$ and whose fundamental groups surject onto surface groups of genus $\geq 2$, although these manifolds admit no non–constant holomorphic maps to compact Riemann surfaces of genus $\geq 2$.

By (the proof of) Corollary 1.66, every finitely presentable group is the fundamental group of some compact complex threefold $Z$ obtained as the twistor space of a suitable anti–self–dual four–manifold $X$. The fibers $F$ of the projection $\pi\colon Z \to X$ are smooth rational curves with normal bundle $\mathcal{O}(1) \oplus \mathcal{O}(1)$, cf. Example 1.64. The exact sequence of sheaves

$$0 \longrightarrow \mathcal{O}_F(-1)^2 \longrightarrow \Omega^1_Z|_F \longrightarrow \Omega^1_F \longrightarrow 0$$

shows that there are no non–zero global holomorphic 1–forms on $Z$, because there are none on the fibers and $\mathcal{O}_F(-1)^2$ has no sections. Thus, no twistor space admits a surjective holomorphic map to a compact Riemann surface of positive genus, no matter what its fundamental group is.

Here is a particularly simple example:

EXAMPLE 2.15. Let $X = S^2 \times C_g$, where $C_g$ is a compact surface of genus $g \geq 2$. A product metric on $X$ of constant curvature metrics with sectional curvatures $K_{S^2} = -K_{C_g}$ is conformally flat. This means that its Weyl tensor vanishes identically; in particular it is anti–self–dual. The twistor space has the same fundamental group as $C_g$, but admits no holomorphic map to $C_g$, for any complex structure on the latter.

In this case, the twistor space is actually diffeomorphic to the Kähler manifold $\mathbb{P}(\mathcal{O}_X \oplus \mathcal{O}_X(K))$, where $X$ has the product Kähler structure.

## 4. Fibering compact complex surfaces

We have just seen that the characterisation of complex manifolds which fiber holomorphically over hyperbolic Riemann surfaces in terms of homomorphisms of the fundamental group to surface goups depends on the Kähler condition, at least if the dimension is at least 3. In the light of the results of Section 3 in Chapter 1, it is perhaps not surprising that this result is true for all compact complex surfaces, without any additional assumption. What is a little surprising, is that although the Castelnuovo–De Franchis Theorem 2.7 (cf. Remark 2.8) and the Siu–Beauville Theorem 2.11 are true for compact complex surfaces, one cannot deduce the latter from the former. Indeed, the intermediate step, Catanese's Corollary 2.9 is false without the Kähler assumption.

EXAMPLE 2.16. Let $X$ be a primary Kodaira surface. This is an elliptic surface over an elliptic curve with $b_1(X) = 3$, $b_2(X) = 4$ and with zero signature. The image of $\Lambda^2 H^1(X, \mathbb{R})$ in $H^2(X, \mathbb{R})$ is an isotropic subspace for the cup–product, as $\Lambda^4 H^1(X, \mathbb{R}) = 0$. But the signature of the cup product on $H^2(X, \mathbb{R})$ is zero, so the maximal dimension of an isotropic subspace is $\frac{1}{2} b_2(X) = 2$. Thus there is an isotropic subspace of dimension at least 2 in $H^1(X, \mathbb{R})$. On the other hand, the first Betti number is too small for $X$ to admit a surjective map to a curve of genus $\geq 2$.

Here is the Siu–Beauville Theorem for compact complex surfaces:

THEOREM 2.17 (Kotschick). *Let $X$ be a compact complex surface and $g \geq 2$ an integer. Then $X$ admits a surjective holomorphic map to some compact Riemann surface of genus $g' \geq g$ having connected fibers if and only if there is a surjective homomorphism $h \colon \pi_1(X) \to \pi_1(C_g)$, with $\pi_1(C_g)$ the fundamental group of a compact Riemann surface of genus $g$.*

PROOF. If $b_1(X)$ is even, then $X$ is Kähler by Theorem 1.28, and so we can apply Theorem 2.11.

If $b_1(X)$ is odd, we appeal to Theorem 1.27. If $h \colon \pi_1(X) \to \pi_1(C_g)$ is surjective, then $b_1(X) \geq 2g + 1$. The assumption $g \geq 2$ implies that $X$ is elliptic. Let $X \to C'$ be an elliptic fibration with connected fibers. Then $b_1(C') \geq b_1(X) - 2 \geq 2g - 1$, and thus $g' = g(C') \geq g$. □

CHAPTER 3

# The de Rham fundamental group

The aim of this chapter is to introduce Sullivan's theory of one–minimal models, to briefly explain how it relates to the de Rham fundamental group, and to show how in the case of a compact Kähler manifold both are determined by the real cohomology algebra. Some features of Kähler groups will be derived from this, and we will present examples of groups which cannot be Kähler because they lack these features.

The key property of compact Kähler manifolds on which all this rests is *formality*, which is a particular type of rigidity over the cohomology algebra. Some such rigidity properties of low–dimensional manifolds are well–known:

- (Radó) Let $X$ be a closed connected surface. Then the Euler characteristic $\chi(X)$ together with the orientability type of $X$ determine its homeomorphism type.
- (Freedman) Let $X$ be an oriented simply–connected closed smooth manifold of dimension 4. Then the cohomology algebra $H^*(X, \mathbb{Z})$ determines the homeomorphism type of $X$.

In the case of compact Kähler manifolds of arbitrary dimension and fundamental group, an example due to Serre shows that the integral cohomology algebra does not determine even the homotopy type. Yet it turns out that the *real* cohomology algebra of such a manifold determines its *real* homotopy type. This is what is known as *formality*, and we will explore some of its consequences for the fundamental group.

Throughout this Chapter we use real coefficients in cohomology and homotopy, as this allows the introduction of Sullivan's minimal models by means of the de Rham complex. Nevertheless, all the results stated are also valid with rational coefficients, or indeed with coefficients in any field of characteristic zero.

This Chapter has benefited greatly from conversations with J. Kollár and V. Navarro Aznar.

## 1. The de Rham fundamental group and the 1–minimal model

Let $X$ be a topological space with finitely presentable fundamental group. The *de Rham fundamental group* of $X$, denoted $\pi_1(X) \otimes \mathbb{R}$, is the $\mathbb{R}$–unipotent completion of the fundamental group of $X$, as described in Appendix A. It is a projective limit of $\mathbb{R}$–unipotent groups $(\pi_1(X)/\pi_1(X)_n) \otimes \mathbb{R}$ and is naturally equivalent to a pro–$\mathbb{R}$–nilpotent Lie algebra, the $\mathbb{R}$–*Malcev algebra* of $\pi_1(X)$, denoted $\mathcal{L}(\pi_1(X), \mathbb{R})$. For the sake of simplicity, we will only deal with real cohomology and real unipotent completions throughout this chapter, and thus in what follows we will sometimes write $H^*(X)$ and $\mathcal{L}\pi_1(X)$, omitting the coefficient ring $\mathbb{R}$.

As explained in Appendix A, the de Rham fundamental group contains the torsion–free nilpotent completion of the fundamental group as a pro–cocompact lattice, and so the two completions are very closely related. The de Rham fundamental group owes its name to the fact that, when $X$ is a smooth manifold, this group can be computed from the de Rham complex. Sullivan's construction of 1–minimal models gives us an algorithm for doing this. In order to describe the algorithm, we need first to define a few concepts.

DEFINITION 3.1. (1) A *commutative differential graded algebra* (CDGA) is a graded algebra $A$ with the following two properties:

- $A$ is graded–commutative, i.e.,
$$y \wedge x = (-1)^{|x||y|} x \wedge y$$
for any two homogeneous elements $x, y \in A$ of degree $|x|, |y|$ respectively;
- there is a boundary operator $d \colon A \to A$ of degree one such that $d^2 = 0$ and
$$d(x \wedge y) = dx \wedge y + (-1)^{|x|} x \wedge dy \ .$$

Morphisms of CDGAs must respect the degree and the boundary operator.

(2) A *quasi–isomorphism* of CDGAs is a CDGA morphism inducing an isomorphism in cohomology.

A quasi–isomorphism does not necessarily have an inverse in the category of CDGAs; the following definition remedies this problem:

DEFINITION 3.2. Two CDGAs $A$ and $B$ are *weakly equivalent* if there is a finite diagram of CDGAs
$$A \to C_1 \leftarrow C_2 \to \cdots \leftarrow B$$
such that all the morphisms are quasi–isomorphisms.

The basic example of a CDGA is the de Rham algebra $\mathcal{E}^*(X)$ of smooth forms on a smooth manifold $X$.

Two other important notions in the CDGA category are:

DEFINITION 3.3. (i) A *base point* of a CDGA $A$ is a CDGA morphism
$$\varepsilon \colon A \longrightarrow \mathbb{R} \ ,$$
where $\mathbb{R}$ is a CDGA with differential $d = 0$.

(ii) A *basepointed homotopy* between two CDGA morphisms $\varphi_0, \varphi_1 \colon A \to B$ is a CDGA morphism
$$\Phi \colon A \longrightarrow \mathbb{R}(t, dt) \otimes B \ ,$$
where $\mathbb{R}(t, dt)$ is the CDGA determined by assigning degree 0 to $t$ and setting the obvious differential, such that given the two base points of $\mathbb{R}(t, dt)$, $\varepsilon_0, \varepsilon_1$, obtained by sending $t$ to 0 and 1 respectively and $dt$ to 0, one has the identities
$$\varphi_0 = (\varepsilon_0 \otimes \mathrm{Id}) \circ \Phi \ , \qquad \varphi_1 = (\varepsilon_1 \otimes \mathrm{Id}) \circ \Phi \ .$$

Again, the motivating examples for these definitions are the notion of a base point of a smooth manifold $x \in X$, with the evaluation map of forms at $x$, and the notion of the codifferential of a homotopy of smooth maps $H \colon [0, 1] \times X \to Y$.

Let $X$ be a smooth manifold, and $\mathcal{E}_X^*$ its de Rham complex. The theory of minimal models developed by Sullivan shows that $\mathcal{E}^*(X)$ has a 1–minimal model. This is a certain free commutative differential graded algebra $M_X(2, 0)$, or simply $M_X$, defined as the limit of an inductive system of CDGAs
$$M_X(1, 1) \hookrightarrow M_X(1, 2) \hookrightarrow M_X(1, 3) \hookrightarrow \ldots \ ,$$

together with a morphism $\rho\colon M_X \to \mathcal{E}^*(X)$ such that in cohomology $\rho^*$ induces isomorphisms in $H^0$ and $H^1$ and a monomorphism $H^2(M_X) \xrightarrow{\rho^*} H^2(\mathcal{E}^*(X))$.

We review the construction of the 1–minimal model up to the second step $M(1,2)$, which will be used to relate $\pi_1(X)$ to the cup products of 1–forms. For a more detailed discussion of 1–minimal models we refer the reader to [56].

Define $M(1,1)$ as the free CDGA $\wedge(V_1^1)$, where $V_1^1$ is the $\mathbb{R}$-vector space $H^1(X,\mathbb{R})$. Every element of $V_1^1$ is defined to have degree one and boundary zero, and the map $\rho\colon M(1,1) \to \mathcal{E}^*(X)$ sends every $x \in V_1^1 = H^1(X,\mathbb{R})$ to its image in a fixed arbitrary linear section $H^1(X,\mathbb{R}) \to$ (cocycles)$^1$ of the projection.

The $(1,2)$-minimal model is defined as an extension of $M(1,1)$: $M(1,2) = \wedge(V_1^1 \oplus V_2^1)$, where $V_2^1$ is the $\mathbb{R}$-vector space $\ker(H^2 M(1,1) \xrightarrow{\rho^*} H^2(X,\mathbb{R}))$. For any $v \in V_2^1$ we define its boundary $dv$ as the element of $\ker H^2 \rho \subset V_1^1 \wedge V_1^1$ defining its cohomology class, and if $dv = \sum x_i y_i \in M(1,1)$, then $\rho(v)$ is a linearly varying primitive of $\sum \rho(x_i)\rho(y_i)$ in $\mathcal{E}^*(X)$.

REMARK 3.4. By definition, $H^2 M(1,1) \cong H^1(X,\mathbb{R}) \wedge H^1(X,\mathbb{R})$, hence there is an isomorphism $V_2^1 \cong \ker(H^1(X,\mathbb{R}) \wedge H^1(X,\mathbb{R}) \xrightarrow{\cup} H^2(X,\mathbb{R}))$.

The subsequent steps $M(1,n)$ are constructed similarly, defining $V_n^1$ as $\ker(H^2 M(1,n-1) \to H^2 \mathcal{E}^*(X))$ and defining $d$ and $\rho$ as for $n=2$. The inductive limit, $M(2,0)$ or $M_X$, is the 1–minimal model of $\mathcal{E}^*(X)$. To achieve functoriality, one must fix a base point for the de Rham algebra and select $\rho$ at every step so that the morphism $\rho\colon M(1,n) \to \mathcal{E}^*(X)$ preserves the base point. Our subsequent use of the 1–minimal model allows us to ignore this issue.

The 1–minimal model is the first step in an inductive system which forms Sullivan's (full) minimal model of a CDGA $A$ and is the simplest CDGA which is weakly equivalent to the original algebra $A$. Some relevant properties of the 1–minimal model are:

PROPOSITION 3.5 (Sullivan). (i) *All connected CDGAs have a 1–minimal model.*
(ii) *After fixing a basepoint, the 1–minimal model of a CDGA is well–defined up to isomorphism.*
(iii) *The 1–minimal model is functorial; i.e., any basepoint-preserving CDGA morphism* $\mathcal{E}^*(Y) \xrightarrow{f} \mathcal{E}^*(X)$ *may be lifted to a morphism* $M_Y \xrightarrow{M(f)} M_X$.
(iv) *Weakly equivalent CDGAs have isomorphic 1–minimal models.*

By (iv) we may compute the 1–minimal model of a manifold $X$ by computing it for any CDGA linked by a chain of quasi–isomorphisms to $\mathcal{E}^*(X)$. Another interesting consequence of this proposition is:

COROLLARY 3.6. *Let* $f\colon X \to Y$ *be a map between smooth manifolds such that the induced map* $f^*$ *on real cohomology is an isomorphism on* $H^0$ *and* $H^1$ *and a monomorphism on* $H^2$. *Then* $f^* \circ \rho_Y \colon M_Y \to \mathcal{E}^*(X)$ *is a 1–minimal model for* $\mathcal{E}^*(X)$.

REMARK 3.7. Example 1.20 shows that for every CW–complex $X$ there exist an aspherical space $Y = K(\pi_1(X), 1)$ and a map $X \to Y$ which satisfy the hypotheses of the Corollary.

We now recall the dualising process between Lie algebras and free commutative differential graded algebras generated by elements of degree one.

Let $L$ be a finite–dimensional $\mathbb{R}$–Lie algebra. Its bracket is a bilinear alternating map
$$[.,.] : L \wedge L \longrightarrow L .$$
Dualising on both sides, the bracket $[.,.]$ has an adjoint map
$$d \colon L^\vee \longrightarrow L^\vee \wedge L^\vee .$$
The map $d$ may be extended as a graded derivation to the free graded algebra $\Lambda L^\vee$, defining the degree of elements in $V^\vee$ to be one. Then the Jacobi identity satisfied by $[.,.]$ dualises as $d^2 = 0$.

Conversely, if $M = \Lambda W$ is a free CDGA and $\deg W = 1$, the differential restricts to a map $d \colon W = M^1 \to M^2 = W \wedge W$, which dualises to a map $[.,.] \colon W^\vee \wedge W^\vee \to W^\vee$, and the fact $d^2 = 0$ in $M$ translates as the Jacobi identity in $W^\vee$.

DEFINITION 3.8. A Lie algebra $L$ and a free CDGA generated by elements of degree one are *dual* when each one yields the other by the above processes.

THEOREM 3.9 (Sullivan (1977)). *Let $X$ be an arc–connected smooth manifold with a finitely presentable fundamental group $\pi_1(X, *)$. The inductive system*
$$M(1,1) \hookrightarrow M(1,2) \hookrightarrow \ldots$$
*formed by the $(1,n)$–minimal models of $X$ and the projective system of real Malcev algebras of the fundamental group*
$$\cdots \to \mathcal{L}_2(\pi_1(X), \mathbb{R}) \to \mathcal{L}_1(\pi_1(X), \mathbb{R})$$
*are dual to each other.*

This theorem has important consequences for our purposes. Foremost is the following duality between the linear spaces $V_n^1$ and the quotients of the lower central series, for whose definition we refer to Appendix A:

COROLLARY 3.10. $V_n^1 \cong ((\pi_1(X)_n / \pi_1(X)_{n+1}) \otimes \mathbb{R})^\vee$

Sullivan's theory of minimal models has two main geometric applications. First, it allows the computation of the Malcev algebra, and thus of the de Rham fundamental group, of many smooth manifolds. Second, for a simply connected manifold, there is a theorem of Sullivan's analogous to Theorem 3.9 stating that the (full) minimal model of the manifold is equivalent to its real Postnikov tower, thereby yielding its real homotopy type. For an introduction to minimal models of simply connected spaces, we refer the reader to [56], and for a unified approach complete with proofs, to [21].

## 2. Formality of compact Kähler manifolds

In the previous Section we described the construction of Sullivan's 1–minimal model and its relation to the de Rham fundamental group in the case of arbitrary smooth manifolds. We shall now study a special property of compact Kähler manifolds with respect to their real homotopy, which is related to Metatheorem 1.2. The abstract property is the following:

DEFINITION 3.11. A smooth manifold $X$ is *formal* if the CDGAs $\mathcal{E}^*(X)$ and $H^*(X, \mathbb{R})$ are weakly equivalent.

This is equivalent to $\mathcal{E}^*(X)$ and $H^*(X,\mathbb{R})$ having isomorphic minimal models. Thus by Sullivan's theory, the real homotopy type of $X$ is determined by its real cohomology algebra.

EXAMPLE 3.12. Formality is a common property among manifolds with a simple cohomology algebra. Some particular examples are:
1. spheres and wedges of spheres,
2. compact connected Lie groups,
3. Eilenberg–Mac Lane spaces $K(\pi, n)$ for $n > 1$,
4. Riemannian symmetric spaces,
5. complements of hyperplane arrangements in $\mathbb{C}^n$.

Another important class of formal spaces is that of compact Kähler manifolds. These are formal as an immediate consequence of Hodge theory, notably of the $dd^c$–Lemma B.7 in Appendix B, and this is the main result of this Chapter.

THEOREM 3.13 (Deligne–Griffiths–Morgan–Sullivan [41]). *Compact Kähler manifolds are formal.*

PROOF. Let $X$ be a compact Kähler manifold. Consider the de Rham complex $\mathcal{E}_X^*$, the subcomplex $\mathcal{E}_X^c$ formed by the $d^c$–closed forms, and the quotient complex $H_{d^c}(X) = \mathcal{E}_X^c / d^c \mathcal{E}_X^*$. They form a diagram

$$(\mathcal{E}_X^*, d) \xleftarrow{i} (\mathcal{E}_X^c, d) \xrightarrow{\rho} (H_{d^c}(X), d) \, .$$

To begin, observe that $d$ induces the zero map in $H_{d^c}(X)$. This follows by applying the $dd^c$–Lemma B.7 to $dy$ for any $y \in \mathcal{E}_X^*$. Thus $(H_{d^c}(X), d) \cong (H_{d^c}^*(X), 0) \cong (H_d^*(X), 0)$, i.e., this is the cohomology complex $H^*(X, \mathbb{R})$.

Next we check that $i$ is a quasi–isomorphism of CDGAs. The induced morphism $i^*$ is surjective: Let $\alpha \in H^*(\mathcal{E}_X, d)$, and choose a representing cocycle $a$. The form $d^c a$ fulfills the hypothesis of the $d^c d$–Lemma, so $d^c a = d d^c w$. The form $y = a + dw$ is another representative for $\alpha$, and $y \in \mathcal{E}_X^c$.

The morphism $i^*$ is injective: Consider a form $y \in \mathcal{E}_X^c$ such that $y = dz$. By the $dd^c$–Lemma, $y = d d^c w$ for some form $w \in \mathcal{E}_X^*$. Thus $y$ admits a primitive $d^c w \in \mathcal{E}_X^c$ and is cohomologically trivial in this complex.

Finally, we prove that $\rho$ is a quasi–isomorphism. The induced cohomology morphism $\rho^*$ is surjective: This follows immediately from its definition as a projection from the $d^c$-cocycle complex $\mathcal{E}_X^c$ to the $d^c$-cohomology complex.

The morphism $\rho^*$ is injective: If $y \in \mathcal{E}_X^c$ is $d$–closed and is in the image of $d^c$, then by the $d^c d$–Lemma $y = d^c dz = d d^c(-z)$, so $y$ is already exact in $\mathcal{E}_X^c$.

Thus we have linked $(\mathcal{E}^*(X), d)$ and $H^*(X, \mathbb{R})$ by a chain of quasi–isomorphisms. Therefore their minimal models are isomorphic, by the definition of formality. □

We shall concentrate on the implications of Theorem 3.13 for the 1–minimal model and the de Rham fundamental group.

COROLLARY 3.14. *The de Rham fundamental group of a compact Kähler manifold is determined by the cup product* $\cup : H^1(X, \mathbb{R}) \otimes H^1(X, \mathbb{R}) \to H^2(X, \mathbb{R})$.

PROOF. The de Rham fundamental group $\pi_1(X) \otimes \mathbb{R}$ is equivalent to the real Malcev algebra $\mathcal{L}(\pi_1(X), \mathbb{R})$, which is dual to the 1–minimal model $M_X$ of $X$.

In the case when $X$ is a compact Kähler manifold, by Theorem 3.13 the 1–minimal model $M_X$ is also the 1–minimal model of the cohomology algebra

$H^*(X,\mathbb{R})$. As may be seen from its construction in the previous Section, this 1–minimal model is determined by the cohomology group $H^1(X,\mathbb{R})$ and the cup product $H^1(X,\mathbb{R}) \otimes H^1(X,\mathbb{R}) \to H^2(X,\mathbb{R})$. □

Topological spaces $X$ for which Corollary 3.14 holds are called 1–formal spaces. An equivalent and more precise definition is:

DEFINITION 3.15. A topological space $X$ is 1–*formal* if there exists a CDGA morphism
$$\rho \colon M_X(2,0) \longrightarrow H^*(X)$$
such that $H^0(\rho)$ and $H^1(\rho)$ are isomorphisms and $H^2(\rho)$ is a monomorphism.

In the case of a formal manifold, the weak equivalence between the CDGAs $\mathcal{E}^*(X)$ and $H^*(X)$ readily implies that $X$ is 1–formal. The 1–formality property actually depends only on the fundamental group; indeed it may be characterised in terms of aspherical spaces: a topological space $X$ is 1–formal if and only if the Eilenberg–Mac Lane space $K(\pi_1(X),1)$ is.

## 3. Applications to the fundamental group and examples

**3.1. Quadratic presentation of the Malcev algebra.** We shall describe the prime consequence of 1–formality for the de Rham fundamental group. This is the existence of a quadratic presentation of the Malcev algebra $\mathcal{L}\pi_1(X)$, which is actually equivalent to 1–formality. It is easy to write down examples of groups that cannot be Kähler because their Malcev algebras do not fulfill this property.

Recall the following concepts of Lie theory. Given a finite–dimensional $\mathbb{R}$–vector space $H$, the *free Lie algebra* spanned by $H$, which we will denote by $L(H)$, is the sub–Lie algebra of the tensor algebra $T(H) = \bigoplus_{n \geq 0} T^n(H) = \bigoplus_{n \geq 0} H^{\otimes n}$ generated by $H$, with the bracket given by
$$[u,v] = u \otimes v - v \otimes u.$$

The free Lie algebra $L(H)$ may be alternatively characterised by a universal property, as the functor $H \mapsto L(H)$ is the left adjoint of the inclusion of $\mathbb{R}$–Lie algebras into $\mathbb{R}$–vector spaces. Another alternative presentation in terms of Malcev algebras is the isomorphism
$$L(H) \cong \mathcal{L}(F_{\dim H}),$$
where $F_{\dim H}$ is the free group of rank $\dim H$. Let us fix some notation:

- The *lower central series* of a Lie algebra will be denoted by $\mathcal{C}^1 L = L, \mathcal{C}^2 L = [L,L],\ldots,\mathcal{C}^n L = [\mathcal{C}^{n-1}L, L],\ldots$.
- The *quadratic elements* of $L(H)$ are the elements of the linear subspace $(L(H) \cap T^2(H)) \cong \wedge^2 H$.
- An ideal $J \subset L(H)$ is *quadratically generated* if it is generated by quadratic elements.
- A *quadratically presented* Lie algebra is the quotient $L(H)/J$ of a free Lie algebra $L(H)$ by a quadratically generated ideal $J$.

It is clear that the class of quadratically presented Lie algebras is very narrow. However, it has been shown by Carlson–Toledo and by S. Chen in [**34**] that this class contains nilpotent algebras of arbitrarily large nilpotency class.

All Lie algebras are quotients of free Lie algebras, but in the case of Malcev algebras such a quotient presentation can be given naturally:

## 3. APPLICATIONS TO THE FUNDAMENTAL GROUP AND EXAMPLES

LEMMA 3.16. *Let $\Gamma$ be a finitely presentable group and $\mathcal{L}\Gamma$ its real Malcev algebra. There is an isomorphism of Lie algebras*
$$\mathcal{L}\Gamma \cong L(H_1(\Gamma, \mathbb{R}))/J,$$
*where $J \subset \mathcal{C}^2 L(H)$, and is a finitely generated ideal.*

For the proof, the reader may consider a morphism from a free group $F_{\dim H_1 \Gamma} \to \Gamma$ inducing an isomorphism on $H_1/_{torsion}$.

Next, following Morgan, we shall give a cohomological characterisation of quadratically presented Malcev algebras. First, let us recall that the tower of $n$–step Malcev algebras
$$\cdots \to \mathcal{L}_n \Gamma \to \cdots \to \mathcal{L}_1 \Gamma$$
or, equivalently, the dual inductive system of minimal CDGAs
$$M_\Gamma(1,1) \hookrightarrow \cdots \hookrightarrow M_\Gamma(1,n) \hookrightarrow \cdots$$
define an inductive system of cohomology maps
$$H^2(\mathcal{L}_n \Gamma) \cong H^2 M_\Gamma(1,n) \xrightarrow{\nu_n^*} H^2 M_\Gamma(2,0) \cong H^2(\mathcal{L}\Gamma).$$

LEMMA 3.17. *Let $\Gamma$ be a finitely presentable group. Then the following are equivalent:*
(i) *its Malcev algebra $\mathcal{L}\Gamma$ admits a quadratic presentation,*
(ii) *the map*
$$H^2(\mathcal{L}_1 \Gamma) \xrightarrow{\nu_1^*} H^2(\mathcal{L}\Gamma)$$
*is surjective,*
(iii) *there is an action of the multiplicative group $\mathbb{R}^*$ by automorphisms on $\mathcal{L}\Gamma$ so that $\lambda \in \mathbb{R}^*$ acts as multiplication by $\lambda$ on $H^1(\mathcal{L}\Gamma)$ and as multiplication by $\lambda^2$ on $H^2(\mathcal{L}\Gamma)$.*

REMARK 3.18. The action in statement (iii) is necessarily by semi–simple automorphisms.

PROOF. Consider the presentation $\mathcal{L}\Gamma \cong L(H)/J$, where $H = H_1(\Gamma, \mathbb{R})$ and $J$ is a finitely generated ideal in $\mathcal{C}^2 L(H)$. This presentation arises from an exact sequence of $L(H)$–modules
$$(5) \qquad 0 \longrightarrow J \longrightarrow L(H) \longrightarrow \mathcal{L}\Gamma \longrightarrow 0.$$
Taking cohomology with coefficients in the trivial module $\mathbb{R}$, and the action of $L(H)$ on the cohomology groups, there is an exact sequence of cohomology groups (see [74])
$$0 \longrightarrow H^1(\mathcal{L}\Gamma) \longrightarrow H^1(L(H)) \longrightarrow H^1(J)^{L(H)} \longrightarrow H^2(\mathcal{L}\Gamma) \longrightarrow H^2(L(H)).$$
As $L(H)$ is a free algebra, $H^2(L(H)) = 0$. Moreover, the fact that $J \subset \mathcal{C}^2 L(H)$ also means that the map $H^1(\mathcal{L}\Gamma) \to H^1(L(H))$ is an isomorphism. The action of $L(H)$ on $H^1(J)$ yields an isomorphism
$$(6) \qquad H^2(\mathcal{L}\Gamma) \cong H^1(J)^{L(H)} \cong (J/[J, L(H)])^\vee.$$

For every $n$–step Malcev algebra one may repeat this reasoning with the presentation
$$0 \longrightarrow J + \mathcal{C}^{n+1} L(H) \longrightarrow L(H) \longrightarrow \mathcal{L}_n \Gamma \longrightarrow 0,$$

and thus obtain an isomorphism
$$H^2(\mathcal{L}_n\Gamma) \cong H^2 M_\Gamma(1,n) \cong (J/([J,L(H)] + \mathcal{C}^{n+2}L(H) \cap J)$$
$$+ \mathcal{C}^{n+1}L(H)/([J,L(H)] \cap \mathcal{C}^{n+1}L(H) + \mathcal{C}^{n+2}L(H)))^\vee.$$

The second term $\left(\mathcal{C}^{n+1}L(H)/([J,L(H)] \cap \mathcal{C}^{n+1}L(H) + \mathcal{C}^{n+2}L(H))\right)^\vee$ lies in the kernel of the cohomology map $H^2(M(1,n)) \to H^2(M(1,n+1))$, hence it has trivial image in $H^2(\mathcal{L}\Gamma)$ and the latter is the inductive limit
$$(J/[J,L(H)])^\vee \cong \varinjlim \left(J/([J,L(H)] + \mathcal{C}^{n+2}L(H) \cap J)\right)^\vee.$$

Thus the morphism $H^2(\mathcal{L}_1\Gamma) \xrightarrow{\nu_1^*} H^2(\mathcal{L}\Gamma)$ is onto if and only if
$$\mathcal{C}^3 L(H) \cap J \subset [J, L(H)],$$
and this inclusion is equivalent to $J$ being generated by the finite–dimensional linear space of quadratic elements $J \cap T^2 H$. This proves the equivalence between conditions (1) and (2).

To prove the equivalence between conditions (1) and (3), first observe that $\mathbb{R}^*$ acts on $\mathcal{L}\Gamma$, with the action on $H^1$ being multiplication by $\lambda$, if and only if $J$ is a homogeneous ideal. Namely the $\mathbb{R}^*$-action on $H$ extends uniquely to the $\mathbb{R}^*$-action on $L(H)$, where the action on homogeneous elements of degree $k$ in $L(H)$ is multiplication by $\lambda^k$. This action is the only possible lifting of each automorphism in $\mathbb{R}^*$ from $L(H)/J$ to $L(H)$. Moreover, this action on $L(H)$ descends to an action on $L(H)/J$ if and only if $J$ is invariant under the action, which is equivalent to the definition of a homogeneous ideal.

Note that this argument shows that any action on $L(H)/J$ which is multiplication by $\lambda$ on $H$ must be by semi–simple automorphisms. In particular, this justifies Remark 3.18.

Now equation (6) shows that the induced action on $H^2(\mathcal{L}\Gamma)$ is multiplication by $\lambda^2$ if and only if
$$[J, L(H)] = \mathcal{C}^3 L(H),$$
i.e., if and only if $\mathcal{L}\Gamma$ is quadratically presented. □

REMARK 3.19. One may check in the same way the following generalisation of the first equivalence in Lemma 3.17: For any $m \geq 2$, $\mathcal{L}\Gamma$ admits a presentation $L(H)/J$ where $J$ has generators in $T^2 H \oplus \cdots \oplus T^m H$ if and only if the morphism $H^2(\mathcal{L}_{m-1}\Gamma) \to H^2(\mathcal{L}\Gamma)$ is onto.

We are now ready to establish the equivalence of the two versions of 1–formality present in the literature:

PROPOSITION 3.20 (Morgan). *Let $X$ be a topological space with a finitely presentable fundamental group. Then $X$ is 1–formal if and only if its Malcev algebra $\mathcal{L}\pi_1(X)$ is quadratically presented.*

PROOF. Due to Lemma 3.17 and the property of the Malcev algebra that the morphism
$$H^2(\mathcal{L}\pi_1(X)) \cong H^2(M_X) \xrightarrow{H^2\rho} H^2(X)$$
is a monomorphism, it suffices to show that $X$ is 1–formal if and only if $\operatorname{Im} H^2\rho = \operatorname{Im} H^2\rho_{(1,1)}$, i.e., that the images in $H^2(X)$ of
$$\rho_{(1,n)}: M(1,n) \longrightarrow \mathcal{E}^*(X),$$

which form an increasing chain of subspaces by the definition of the minimal CDGAs $M(1,n)$, stabilise at the step $n=1$.

If $X$ is 1–formal, then its 1–minimal model is isomorphic to that of the algebra $H^*(X)$. We can build a 1–minimal model for $H^*(X)$ such that, if $M(1,n) = \Lambda(V_1^1 \oplus \cdots \oplus V_n^1)$, then the morphism $\rho: M(2,0) \to H^*(X)$ verifies that $\rho_{|V_m^1} = 0$ for $m \geq 2$. This may be done as follows:

Let $M(1,1) = \Lambda(V_1^1)$ with $\rho_{(1,1)}: V_1^1 \xrightarrow{\cong} H^1(X)$ be the first step of the minimal model. The following steps are defined by adjoining spaces

$$V_n^1 = \ker\left(H^2(M(1,n-1)) \xrightarrow{\rho_{(1,n-1)}} H^2(X)\right),$$

and we can define $\rho_{(1,n)}$ over $V_n^1$ as any linear map such that $d\rho_{(1,n)}v = \rho_{(1,n-1)}dv = 0 \in H^2(X)$. Therefore we may set $\rho_{|V_n^1} = 0$, for all $n > 1$.

The 1–minimal model of $H^*(X)$ that we have just described obviously verifies that $\operatorname{Im} H^2(M(1,1)) = \operatorname{Im} H^2(M(2,0)) \subset H^2(X)$, therefore the Malcev algebra $\mathcal{L}\pi_1(X)$ admits a quadratic presentation.

Conversely, assume that $\mathcal{L}\pi_1(X)$ admits a quadratic presentation. Lemma 3.17 implies that there is an action by Lie algebra automorphisms of $\mathbb{R}^*$ on $\mathcal{L}\Gamma$, and consequently an action by CDGA automorphisms on $M(2,0)$. By Remark 3.18 these automorphisms are semi–simple. This action is equivalent to a grading of $M(2,0)$, namely $M(2,0) = \bigoplus_{k \geq 0} M(2,0)(k)$, where $M(2,0)(k)$ is the subspace of $M(2,0)$ where all $\lambda \in \mathbb{R}^*$ act as $\lambda^k$. This gives a *weight filtration* $W_\bullet$ on $M(2,0)$, defined by

$$W_n = \bigoplus_{k \leq n} M(2,0)(k).$$

This filtration is determined by the facts that it is multiplicative and that the homogeneous elements $v \in V_n^1$ have weight $n$. It is *strongly graded* as in §4,5 of [**38**] and has the property that $H^2(M(2,0))$ is of pure weight 2. This allows the definition of a CDGA morphism

$$\varphi: M_X(2,0) \longrightarrow H^*(X)$$

defined by setting

$$\varphi_{|V_1^1}: V_1^1 \xrightarrow{\cong} H^1(X)$$
$$\varphi_{|V_n^1}: V_n^1 \xrightarrow{0} 0 \in H^1(X) \qquad \text{for } n \geq 2.$$

This is well–defined because for $v \in V_m^1$, with $m \geq 3$, $d\varphi(v) = 0$ and $\varphi(dv) = 0$ as, due to the weight filtration, the monomials in $dv$ contain a factor in $V_k^1$ with $k \geq 2$, while for $v \in V_1^1, V_2^1$ the identity $d\varphi(v) = \varphi(dv) = 0$ is a consequence of the defining properties of $V_1^1, V_2^1$.

By construction, the morphism $\varphi: M_X(2,0) \to H^*(X)$ induces an isomorphism on $H^0$ and $H^1$ and an injection of the subspace $\operatorname{Im} H^2(M(1,1)) \subset H^2(M(2,0))$. As this subspace is the full group $H^2(M(2,0))$, we reach the conclusion that $H^2\varphi$ is a monomorphism, thus $X$ must be 1–formal. $\square$

REMARK 3.21. In the case of compact Kähler manifolds the filtration $W_\bullet$ is indeed the weight filtration of a mixed Hodge structure in the 1–minimal model; see Section 6 for more information.

REMARK 3.22. The property of a minimal model over a field of characteristic zero to have an action of the multiplicative group inducing given weights on cohomology is independent of the field. This is how Sullivan deduces, in §12 of [**130**], that formality is independent of the field; see also Section 6 below.

Proposition 3.20 implies a restrictive necessary condition for a group to be Kähler.

EXAMPLE 3.23. The group $\Gamma = \langle x, y, z, t \,|\, [x,y][z,t], [[[[y,x],x],x],y]\rangle$, all of whose Massey triple products are zero (cf. subsection 3.3 below), cannot be Kähler, because its Malcev algebra does not admit a quadratic presentation.

Another consequence of the quadratic presentation of the Malcev algebra, originally observed by Morgan in the case of Kähler groups, is:

COROLLARY 3.24. *Let $\Gamma$ be a finitely presentable group such that its Malcev algebra is quadratically presented. Its $n$–step Malcev algebras $\mathcal{L}_n(\Gamma, \mathbb{R})$ are isomorphic to the graded Lie algebras induced by the group bracket $gr_n\Gamma \otimes \mathbb{R} = \oplus_{k=1}^n \Gamma_k/\Gamma_{k+1} \otimes \mathbb{R}$.*

PROOF. The weight filtration $W_\bullet$ and its associated $\mathbb{R}^*$–action induces a canonical splitting of the algebras $\mathcal{L}_n(\Gamma, \mathbb{R})$, which is respected by the Lie bracket. The filtration induced by weight coincides with that induced by the lower central series on $\mathcal{L}_n\Gamma$, and the graded Lie algebra induced by the lower central series in $\mathcal{L}_n\Gamma$ is isomorphic to $gr_n\Gamma \otimes \mathbb{R}$ (see the Appendix to [**103**]). □

**3.2. Groups with free Malcev completions.** As was remarked in Example 1.19, free groups are not Kähler for elementary reasons. More generally, we can now show that groups with a free Malcev algebra, or even those "free up to order two brackets", cannot be Kähler.

We will use here the 2–*step nilpotent de Rham fundamental group* $(\pi_1(X)/\pi_1(X)_3) \otimes \mathbb{R}$, or equivalently the 2–*step nilpotent Malcev algebra* $\mathcal{L}_2(\pi_1(X), \mathbb{R})$, which is isomorphic, though not canonically so, to $\operatorname{Gr} \mathcal{L}_2(\pi_1(X), \mathbb{R}) = (\pi_1(X)/\pi_1(X)_2 \otimes \mathbb{R}) \oplus (\pi_1(X)_2/\pi_1(X)_3 \otimes \mathbb{R})$. This is the simplest non–Abelian quotient after $H^1(X, \mathbb{R})$.

PROPOSITION 3.25. *If $\mathcal{L}_2\Gamma \cong \mathcal{L}_2 F_n$ for some free group $F_n$, then $\Gamma$ cannot be a Kähler group.*

PROOF. If $\mathcal{L}_2\Gamma \cong \mathcal{L}_2 F_n$, then $\dim \Gamma/\Gamma_2 \otimes \mathbb{R} = n$, and $\dim \Gamma_2/\Gamma_3 \otimes \mathbb{R} = \dim(F_n)_2/(F_n)_3 \otimes \mathbb{R} = \binom{n}{2}$. Thus if $\Gamma = \pi_1(X)$ with $X$ compact Kähler, by the formality of $X$ and the duality of the de Rham fundamental group with the 1–minimal model, we would have that

$$\dim H^1(X, \mathbb{R}) = \dim V_1^1 = \dim \Gamma/\Gamma_2 \otimes \mathbb{R} = n$$

and

$$\dim \ker(H^1(X, \mathbb{R}) \wedge H^1(X, \mathbb{R}) \xrightarrow{\cup} H^2(X, \mathbb{R})) = \dim V_2^1 = \dim \Gamma_2/\Gamma_3 \otimes \mathbb{R} = \binom{n}{2}.$$

But the first equality implies that $\dim H^1(X, \mathbb{R}) \wedge H^1(X, \mathbb{R}) = \binom{n}{2}$, so in fact, all exterior products of 1–forms on $X$ would be exact. This is not possible for $X$ compact Kähler, because of the Hard Lefschetz Theorem B.5. Hence the statement. □

EXAMPLE 3.26 (1–relator groups). If $\Gamma$ is a Kähler group admitting a presentation with only one relation, $\Gamma = \langle x_1, \ldots, x_n \mid r \rangle$, then either $n = 1$ and $\Gamma \cong \mathbb{Z}/m\mathbb{Z}$, or $r$ must lie in $F\{x_1, \ldots, x_n\}_2$, otherwise $\mathcal{L}_2\Gamma \cong \mathcal{L}_2 F_{n-1}$ or $\mathcal{L}_2 F_n$.

For instance, the groups $\langle x, y, z \mid xyxzxzxy \rangle$, or $\langle x, y \mid [[x,y], y] \rangle$ cannot be Kähler.

EXAMPLE 3.27 (Generic groups with few defining relations). Let $\Gamma$ be a group admitting a presentation with $n$ generators $x_1, \ldots, x_n$ and $s < n$ defining relations $r_1, \ldots, r_s$, such that their images in the Abelianised group $\mathbb{Z}\bar{x}_1 \oplus \ldots \mathbb{Z}\bar{x}_n \cong \mathbb{Z}^n$ are linearly independent. Then $\Gamma$ cannot be Kähler, as $\mathcal{L}_2\Gamma \cong \mathcal{L}_2 F_{n-s}$. This is the generic case among presentations with fewer relations than generators.

For instance, the group $\Gamma = \langle x_1, \ldots, x_5 \mid x_1^2 x_2^2 x_3^2, x_2^2 x_3^2 x_4^2, x_3^2 x_4^2 x_5^2 \rangle$ cannot be Kähler.

**3.3. Massey products and Heisenberg groups.** We now define Massey triple products. These are cohomological operations which, in the case of 1–forms or of spherical cohomology classes in general, are dual to the group bracket, respectively the Whitehead bracket, of representing homotopy classes of loops or spheres.

Let $\alpha, \beta, \gamma \in H^*(X, \mathbb{R})$, of degrees $p, q, r$ respectively, such that $\alpha \cup \beta = 0$, $\beta \cup \gamma = 0$. Choose corresponding cocycles $a, b, c$, and select primitive cochains $f, g$ such that $df = a \cup b$, $dg = b \cup c$. We define the *Massey triple product* $\langle \alpha, \beta, \gamma \rangle$ as the class of $f \cup c + (-1)^{p-1} a \cup g$ in $H^{p+q+r-1}(X)/(\alpha \cup H^{q+r-1}(X) + \gamma \cup H^{p+q-1}(X))$. One can check that this is well–defined in the quotient, although it would not be well–defined as a cohomology class. The Massey triple product may actually be defined in any CDGA as above, and its functoriality follows from the definition. A consequence of its naturality is the following

LEMMA 3.28. *Let $A$ and $B$ be weakly equivalent algebras. The isomorphism $H^*A \cong H^*B$ preserves Massey triple products.*

In the case of compact Kähler manifolds, formality together with the above Lemma allow us to compute Massey triple products rather easily.

PROPOSITION 3.29. *All Massey triple products on a compact Kähler manifold are zero.*

PROOF. Let $X$ be compact Kähler. By the formality of $X$, the algebras $\mathcal{E}^*(X)$ and $H^*(X, \mathbb{R})$ are weakly equivalent, so we can compute Massey triple products in $H^*(X, \mathbb{R})$. The differential is zero by definition, so all Massey products will be zero. $\square$

From Proposition 3.29 we derive a restriction on Kähler groups. It can be seen directly as a consequence of the quadratic presentation of its Malcev algebra, which implies that all types of Massey products of 1–forms in a Kähler group are zero:

COROLLARY 3.30. *Let $\Gamma$ be a Kähler group. Then all Massey triple products of classes of $H^1(\Gamma)$ must be zero.*

PROOF. This follows from Proposition 3.29 and from the fact that if $\Gamma = \pi_1(X)$, there is a map $c \colon X \to K(\Gamma, 1)$ inducing an isomorphism of fundamental groups, an isomorphism on $H^0$ and $H^1$, and a monomorphism $H^2(\Gamma) \to H^2(X, \mathbb{R})$, cf. Example 1.20. Therefore, for Massey products of 1–classes to be zero in a quotient of $H^2(X, \mathbb{R})$, they must be zero in the corresponding quotient of $H^2(\Gamma)$. $\square$

EXAMPLE 3.31 (Serre). The Heisenberg group $\mathcal{H}_3(\mathbb{Z})$ is the group of matrices

$$\mathcal{H}_3(\mathbb{Z}) = \left\{ \begin{pmatrix} 1 & x & z \\ 0 & 1 & y \\ 0 & 0 & 1 \end{pmatrix} \in GL(3, \mathbb{Z}) \right\}.$$

This group is not Kähler, because its cohomology contains nontrivial Massey products.

To check this, let us first observe that $\mathcal{H}_3$ has a dimension 3 Malcev algebra $\langle X, Y, Z \mid [X,Y] = Z, [X,Z] = 0, [Y,Z] = 0 \rangle$. Dualisation of this nilpotent Lie algebra yields the 1–minimal model of any topological space having $\pi_1(X) \cong \mathcal{H}_3$, which is

$$M = \wedge(x,y,z) \qquad \deg x,y,z = 1 \qquad dx = 0,\ dy = 0,\ dz = xy\ .$$

The Massey triple product $\langle x, x, y \rangle$ is well–defined in $H^2(\mathcal{H}_3)$, and it is $xz$, which is a non–zero cohomology class.

EXAMPLE 3.32. Consider the Heisenberg group of Gaussian integers

$$\mathcal{H}_3(\mathbb{Z} \oplus \mathbb{Z}i) = \left\{ \begin{pmatrix} 1 & x & z \\ 0 & 1 & y \\ 0 & 0 & 1 \end{pmatrix} \in GL(3, \mathbb{Z} \oplus \mathbb{Z}i) \right\}.$$

As in the case of the integral Heisenberg group $\mathcal{H}_3(\mathbb{Z})$, we can check that there are nontrivial Massey triple products of 1–forms, and therefore $\mathcal{H}_3(\mathbb{Z} \oplus \mathbb{Z}i)$ is not a Kähler group.

## 4. The Albanese map and the de Rham fundamental group

In this section we describe how the de Rham fundamental group of a compact Kähler manifold is determined by that of its Albanese image. We then discuss some consequences that follow from this combined with knowledge of the structure of the Albanese map[1].

Recall from Section 2 in Chapter 2 that for a compact Kähler manifold $X$, $\alpha_X \colon X \longrightarrow Alb(X)$ denotes its Albanese map. Let $Y = \alpha_X(X)$ be its image, which may be singular. We consider a desingularisation $\varepsilon \colon \tilde{Y} \to Y$, and a desingularisation $\tilde{X}$ of the pullback of $\alpha_X$:

$$\begin{array}{ccc} \tilde{X} & \xrightarrow{\tilde{\alpha}_X} & \tilde{Y} \\ \varepsilon_X \downarrow & & \downarrow \varepsilon \\ X & \xrightarrow{\alpha_X} & Y \end{array}$$

It is clear that the manifold $\tilde{X}$ is also compact Kähler and that the map $\varepsilon_X$ is a birational morphism and thus induces an isomorphism of fundamental groups $\varepsilon_{X*} \colon \pi_1(\tilde{X}) \longrightarrow \pi_1(X)$.

We will call the map $\tilde{\alpha}_X \colon \tilde{X} \longrightarrow \tilde{Y}$ a *smoothing of the Albanese map* of X. The properties of the original Albanese map $\alpha_X$ relate $X$, $\tilde{X}$ and $\tilde{Y}$ as follows:

PROPOSITION 3.33 (Campana). *Let $X$ be a compact Kähler manifold and $\tilde{\alpha}_X \colon \tilde{X} \longrightarrow \tilde{Y}$ a smoothing of its Albanese map. Then $\varepsilon_X$ and $\tilde{\alpha}_X$ induce an isomorphism $\pi_1(X) \otimes \mathbb{R} \xrightarrow{\cong} \pi_1(\tilde{Y}) \otimes \mathbb{R}$.*

---

[1]This section is based in part on a lecture of F. Campana.

## 4. THE ALBANESE MAP AND THE DE RHAM FUNDAMENTAL GROUP

PROOF. As $\varepsilon_X$ induces an isomorphism of fundamental groups, $\varepsilon_X^* \colon H^1(X) \longrightarrow H^1(\tilde{X})$ is also an isomorphism. This implies that $Alb(X)$ is the Albanese torus of $\tilde{X}$ and $\alpha_X \circ \varepsilon_X = \varepsilon \circ \tilde{\alpha}_X$ its Albanese map. As a consequence, $\tilde{\alpha}_X^* \colon H^1(\tilde{Y}) \longrightarrow H^1(\tilde{X})$ is surjective. As $\alpha_X$ itself is also surjective, $\tilde{\alpha}_X^*$ is also injective for $H^*$. Therefore $\tilde{\alpha}_X$ induces an isomorphism $H^1(\tilde{Y}) \cong H^1(\tilde{X})$ and an injection $H^2(\tilde{Y}) \hookrightarrow H^2(\tilde{X})$. Thus, by the universality of the 1–minimal model, $\tilde{\alpha}_X$ induces an isomorphism $M_{\tilde{Y}}(2,0) \cong M_{\tilde{X}}(2,0)$. Dualising, we obtain our statement for the Malcev algebras $\mathcal{L}(\pi_1 \tilde{X})$ and $\mathcal{L}(\pi_1 \tilde{Y})$. The categorical equivalence between real Malcev algebras and de Rham fundamental groups, and the fact that $\varepsilon_X$ induces an isomorphism of fundamental groups $\pi_1(\tilde{X}) \cong \pi_1(X)$, complete the proof. □

Thus the study of de Rham fundamental groups, or equivalently Malcev completions of fundamental groups, of compact Kähler manifolds may be reduced to the study of smoothings of its Albanese images. This is particularly convenient in the following cases:

COROLLARY 3.34 (Campana). *Let $X$ be a compact Kähler manifold with surjective Albanese map $\alpha_X \colon X \longrightarrow Alb(X)$. Then the Albanese map induces an isomorphism of de Rham fundamental groups*

$$\alpha_{X*} \colon \pi_1(X) \otimes \mathbb{R} \xrightarrow{\cong} \pi_1(Alb(X)) \otimes \mathbb{R} \cong \mathbb{R}^{b_1(X)}.$$

PROOF. As $\alpha_X$ is surjective, its image is smooth, and so $\tilde{Y} = Alb(X)$. □

EXAMPLE 3.35. Some examples of Kähler manifolds with surjective Albanese map are: manifolds with Kodaira dimension $\kappa(X) = 0$, manifolds with algebraic dimension $a(X) = 0$ and manifolds with first Betti number $b_1(X) = 0$ or 2.

The image of the Albanese maps of a compact Kähler manifold $X$ is a subvariety of the Albanese torus $Alb(X)$. The following theorem on the structure of subvarieties of complex tori is very useful in the study of their de Rham fundamental groups.

THEOREM 3.36 ([**134**], Theorem 10.9). *Let $Y$ be a subvariety of a complex torus $T$. Then there exists a complex subtorus $A_1 \subset T$ such that $A_2 = T/A_1$ is an Abelian variety, and a projective subvariety $W \subset A_2$ such that:*
  (i) *the natural projection $\pi \colon T \to A_2$ satisfies $Y = \pi^{-1}(W)$, and*
  (ii) *there is an equality of Kodaira dimensions*

$$\kappa(W) = \dim(W) = \kappa(Y).$$

Kollár pointed out to us that this can be reformulated in the following useful form:

COROLLARY 3.37. *Let $Y$ be a subvariety of a complex torus $T$. Then*
  (i) *there are a projective variety $W$ with dimension $\dim W = \kappa(Y)$, and a subtorus $A_1 \subset T$ such that there is a holomorphic map $Y \to W$ and a diffeomorphism over $W$*

$$Y \cong A_1 \times W ;$$

  (ii) *the subvariety $Y$ admits a desingularisation $\tilde{Y}$, together with a holomorphic map $\tilde{Y} \to \tilde{W}$ to a projective desingularisation of $W$, such that $\tilde{Y}$ is diffeomorphic to $A_1 \times \tilde{W}$.*

PROOF. Real tori are semi–simple. Thus, considering the real torus underlying $T$ and its subtorus $A_1$ in Theorem 3.36, there is a diffeomorphism
$$f \colon T \xrightarrow{\cong} A_1 \times A_2$$
with the inclusion $A_1 \hookrightarrow T$ as a factor. Thus by Theorem 3.36 the map $f$ restricts to the desired diffeomorphism $Y \cong A_1 \times W$.

Consider now a projective desingularisation $\tilde{W} \to W$ and the pullback $\tilde{Y}$ in the diagram of holomorphic maps
$$\begin{array}{ccc} \tilde{Y} & \longrightarrow & T \\ \downarrow & & \downarrow \\ \tilde{W} & \longrightarrow & A_2 \end{array}$$
The holomorphic map $\tilde{Y} \to \tilde{W}$ is smooth by base change, and $\tilde{W}$ is smooth. Thus $\tilde{Y}$ is a desingularisation of $Y$. Moreover, $T$ is diffeomorphic over $A_2$ to the trivial family $A_1 \times A_2$, and so $\tilde{Y}$ is diffeomorphic over $\tilde{W}$ to the trivial family $A_1 \times \tilde{W}$. □

Here is another application of Theorem 3.36 to the de Rham fundamental groups of compact Kähler manifolds:

COROLLARY 3.38 (Amorós, Campana). *Let $X$ be a compact Kähler manifold with Kodaira dimension $\kappa(X) = 1$. Then there is a noncanonically split exact sequence*
$$1 \longrightarrow \mathbb{R}^{b_1(X)-2g} \longrightarrow \pi_1(X) \otimes \mathbb{R} \longrightarrow \pi_1(C_g) \otimes \mathbb{R} \longrightarrow 1 \;,$$
*where $C_g$ is a compact curve of genus $g \geq 0$.*

PROOF. The Kodaira dimension of the Albanese image $Y$ of $X$ satisfies the inequality
$$\kappa(Y) \leq \kappa(X) \;.$$
The fact that $Y$ is contained in a complex torus rules out the possibility that $\kappa(Y) = -\infty$. Moreover, if $\kappa(Y) = 0$, the submanifold $Y$ must be a translation of a complex subtorus of $Alb(X)$ because of Theorem 3.36 and must generate $Alb(X)$ because it is the Albanese image of $X$. Therefore $Y = Alb(X)$ and $\pi_1(X) \otimes \mathbb{R} \cong \mathbb{R}^{b_1(X)}$.

If $\kappa(Y) = 1$, by Corollary 3.37 there is a diffeomorphism $\tilde{Y} \cong A_1 \times C_g$ for some smooth compact curve $C_g$, and thus the split exact sequence follows from Proposition 3.33. □

REMARK 3.39. Let us note in addition to the previous proof that when the Albanese image has Kodaira dimension $\kappa(Y) = 1$, then the curve $C_g$ has genus $g \geq 2$, and $X$ is a fibered Kähler manifold.

Moreover, if the Albanese image $Y$ has dimension one, then it must be a smooth curve, the fibers of the Albanese map will be connected ([**134**], Proposition 9.19), and the isomorphism of de Rham fundamental groups comes from an isomorphism of torsion–free nilpotent completions $\pi_1(X)_0^{nilp} \cong \pi_1(C_g)_0^{nilp}$. It is shown in [**54**] that this is the case for Kähler groups admitting a presentation with $n$ generators and $s$ relations where $s \leq n - 2$.

As has been remarked in subsection 2.3 of Chapter 1, an important open question is whether all Kähler groups are fundamental groups of complex projective manifolds. The following Corollary, based on Theorem 3.36, clarifies the question at the de Rham level:

COROLLARY 3.40 (Amorós). *Every de Rham fundamental group of a compact Kähler manifold is the de Rham fundamental group of a complex projective manifold.*

PROOF. By Proposition 3.33, for every Kähler group $\Gamma = \pi_1(X)$, one has an isomorphism of de Rham fundamental groups $\pi_1(X) \otimes \mathbb{R} \cong \pi_1(\tilde{Y})$, where $\tilde{Y}$ is a smoothing of the Albanese image. Moreover, Corollary 3.37 shows that one may choose $\tilde{Y}$ diffeomorphic to $A_1 \times \tilde{W}$, where $A_1$ is a torus and $\tilde{W}$ a projective manifold. Therefore
$$\pi_1(X) \otimes \mathbb{R} \cong \mathbb{R}^{2\dim A_1} \times \pi_1(\tilde{W}) \otimes \mathbb{R} \ .$$
The latter group is the de Rham fundamental group of $A \times \tilde{W}$, with $A$ any Abelian variety of the same rank as $A_1$. □

REMARK 3.41. As we remarked at the beginning of this Chapter, all the results in this section are valid for rational coefficients in cohomology algebras, de Rham groups and Malcev algebras. Moreover, by taking the Stein factorisation of the Albanese map, one can extend Corollary 3.40 and show that all torsion–free nilpotent completions of Kähler groups are torsion–free nilpotent completions of fundamental groups of projective manifolds.

## 5. Non–fibered Kähler groups

Recall from Section 3 in Chapter 2 that Kähler groups are divided into *fibered* and *non–fibered* groups, depending on whether or not they surject onto surface groups of genus $\geq 2$. By Theorem 2.11, a group is fibered if and only if every compact Kähler manifold with that group as fundamental group fibers over a compact Riemann surface of genus $\geq 2$.

We now study the cup products of 1–forms in the case of non–fibered Kähler groups. This is based on the classical Castelnuovo–de Franchis Theorem 2.7. This theorem, together with the conic structure of the set of products in $H^{2,0}(X)$, yields the following corollary:

LEMMA 3.42. *If $X$ is a nonfibered compact Kähler manifold, then*
$$\dim Im \cup \colon H^{1,0}(X) \wedge H^{1,0}(X) \to H^{2,0}(X) \geq 2\dim H^{1,0}(X) - 3 \ .$$

This gives a bound for the products of holomorphic 1–forms and, by conjugation, of anti–holomorphic 1–forms. The dimension of products of holomorphic with anti–holomorphic 1–forms can also be bounded below by applying Castelnuovo–de Franchis (see [**8**], [**2**]). The result is:

PROPOSITION 3.43. *Let $X$ be a non–fibered compact Kähler manifold. Then*
$$\dim \left( Im \cup \colon H^{1,0}(X) \otimes H^{0,1}(X) \longrightarrow H^{1,1}(X) \right) \geq 2 \dim H^{1,0}(X) - 1 \ .$$

We now have all the required pieces in place to bound the second Betti number and the rank of the lower central series quotient $\pi_1(X)_2/\pi_1(X)_3 \otimes \mathbb{R}$ for non–fibered Kähler groups.

PROPOSITION 3.44 ([**2**]). *Let $X$ be a non–fibered compact Kähler manifold with irregularity $q = \frac{1}{2} b_1(X)$.*
  (i) *If $q = 0$ or $1$, then $b_2(\pi_1(X)) \geq 1$ and $\dim(\pi_1(X)_2/\pi_1(X)_3) \otimes \mathbb{R} = 0$.*
  (ii) *If $q \geq 2$, then $b_2(\pi_1(X)) \geq 6q - 7$ and $\dim(\pi_1(X)_2/\pi_1(X)_3) \otimes \mathbb{R} \leq 2q^2 - 7q + 7$.*

PROOF. We have seen that

$$\dim(\pi_1(X)_2/\pi_1(X)_3) \otimes \mathbb{R} = \dim\ker(H^1(X,\mathbb{R}) \wedge H^1(X,\mathbb{R}) \xrightarrow{\cup} H^2(X,\mathbb{R}))$$
$$= \dim(H^1(X,\mathbb{R}) \wedge H^1(X,\mathbb{R}))$$
$$\quad - \dim\operatorname{Im}\left(\cup\colon H^1(X,\mathbb{R}) \wedge H^1(X,\mathbb{R}) \to H^2(X,\mathbb{R})\right)$$
$$= \frac{2q(2q-1)}{2} - \dim\operatorname{Im} \cup \, .$$

For $q \geq 2$, we break $H^1(X,\mathbb{C})$ into its Hodge components. By the above results $\dim\left(\operatorname{Im} H^{1,0}(X) \wedge H^{1,0}(X) \longrightarrow H^{2,0}(X)\right) \geq 2q - 3$. By conjugation, the same holds for $H^{0,1}(X) \wedge H^{0,1}(X) \longrightarrow H^{0,2}(X)$. For holomorphic–antiholomorphic products we have found the inequality $\dim\left(H^{1,0}(X) \wedge H^{0,1}(X) \longrightarrow H^{1,1}(X)\right) \geq 2q - 1$. The claim for the second Betti number now follows by adding the bounds. Subtracting the bound for $\dim\operatorname{Im}\left(H^1(X) \wedge H^1(X) \xrightarrow{\cup} H^2(X)\right)$ from that for $\dim H^1(X) \wedge H^1(X) = \binom{2q}{2}$ yields our bound on $(\pi_1(X)_2/\pi_1(X)_3) \otimes \mathbb{R}$. □

Proposition 3.44 implies in particular that non–fibered Kähler groups need many defining relations. The effect of defining relations on $\dim\Gamma/\Gamma_2 \otimes \mathbb{R}$, $\dim\Gamma_2/\Gamma_3 \otimes \mathbb{R}$ can be estimated, and thus Proposition 3.44 produces the following bounds:

COROLLARY 3.45 ([2]). *Let $\Gamma = \langle x_1, \ldots, x_n \mid r_1, \ldots, r_s \rangle$ be a finite presentation of the group $\Gamma$. If $\Gamma = \pi_1(X)$, with $X$ a non–fibered compact Kähler manifold with irregularity $q = \frac{1}{2} b_1(X)$, then the total number of defining relations must satisfy the following:*
  (i) *if $q = 0$, then $s \geq n$;*
  (ii) *if $q = 1$, then $s \geq n - 1$;*
  (iii) *if $q \geq 2$, then $s \geq n + 4q - 7$.*

EXAMPLE 3.46. Consider $\Gamma = \langle x_1, \ldots, x_{2q} \mid w_1, \ldots, w_s \rangle$ with $w_1, \ldots, w_s \in \langle x_1, \ldots, x_{2q} \rangle_2$. This can only be non–fibered Kähler if $s \geq 6q - 7$ for $q \geq 2$, and $s \geq 1$ for $q = 1$. The same bounds hold for its Malcev algebra.

EXAMPLE 3.47. Consider the group

$$\Gamma = \langle x_1, \ldots, x_5 \mid x_1^2 x_2^{-2} x_4^2, (x_1, x_2), (x_2, x_3), (x_3, x_4), (x_4, x_5) \rangle \, .$$

In this case $n = 5$, $k = 1$, $q = 2$ as $\operatorname{Im} d_0 = \langle 2\bar{x}_1 - 2\bar{x}_2 + 2\bar{x}_4 \rangle$, and $s = 5 < n + 4q - 7 = 6$. Therefore $\Gamma$ cannot be non–fibered Kähler. The group $\Gamma$ cannot map onto a surface group $\pi_1(C_g)$, with $C_g$ a smooth projective curve of genus $g \geq 2$ because $\dim\Gamma_2/\Gamma_3 \otimes \mathbb{R} = 2$, $\dim(\pi_1(C_g)_2/\pi_1(C_g)_3) \otimes \mathbb{R} = \frac{2g(2g-1)}{2} - 1 \geq 5$, so we reach the conclusion that $\Gamma$ cannot be Kähler.

EXAMPLE 3.48. (Groups of planar hyperplane arrangements are not non–fibered Kähler (cf. [100])) Let $\mathcal{A} = \{H_1, \ldots, H_n\}$ be a planar hyperplane arrangement, i.e., a finite set of hyperplanes in $\mathbb{C}^2$, and let $\alpha_j = 0$ be a defining linear equation for every line $H_j$. The complement of the lines is a smooth complex manifold $M(\mathcal{A}) = \mathbb{C}^2 \setminus (H_1 \cup \cdots \cup H_n)$, and its integral cohomology algebra was shown by Brieskorn to be

$$H^*(M;\mathbb{Z}) \cong \langle \frac{1}{2\pi i} \frac{d\alpha_1}{\alpha_1}, \ldots, \frac{1}{2\pi i} \frac{d\alpha_n}{\alpha_n} \rangle \subset \mathcal{E}^*_{\mathbb{C}}(M) \, ,$$

that is, the subalgebra of the complex–valued de Rham complex of $M$ generated by the forms $\omega_j = \frac{1}{2\pi i} \frac{d\alpha_j}{\alpha_j}$.

The above inclusion induces a weak equivalence between the cohomology algebra $H^*(M,\mathbb{R})$ and the de Rham complex of $M$. Therefore the space $M$ is formal, and all the Massey triple products of 1–forms in its cohomology are zero.

Brieskorn's explicit computation of the cohomology of $M$ allows us to present bases for $H^1(M,\mathbb{R})$, $H^2(M.\mathbb{R})$ (see [**100**], Example 7.4):

$$H^1(M,\mathbb{R}) = \langle \omega_1, \ldots, \omega_n \rangle \cong \mathbb{R}^n$$
$$H^2(M,\mathbb{R}) = \mathrm{Im}\left(\cup : H^1(M) \wedge H^1(M) \to H^2(M)\right)$$
$$= \langle \omega_1 \wedge \omega_n, \omega_2 \wedge \omega_n, \ldots, \omega_{n-1} \wedge \omega_n \rangle \cong \mathbb{R}^{n-1}.$$

Therefore, if $n \geq 3$ then $n - 1 \leq 3n - 7$, and by Proposition 3.44 the fundamental group $\Gamma = \pi_1(M)$ cannot be non–fibered Kähler.

On the other hand, in the case $n = 2$, the line arrangement $\{x = 0, y = 0\} \subset \mathbb{C}^2$ yields $\pi_1(M) \cong \mathbb{Z}^2$, which is non–fibered Kähler.

## 6. Mixed Hodge structures on the de Rham fundamental group

We will briefly describe the mixed Hodge structure on the de Rham fundamental group of compact Kähler manifolds due to Morgan [**94**] and a restriction it imposes on Kähler groups. This theory amounts to defining a mixed Hodge structure on the unipotent representations of a Kähler group, and it has recently been complemented by the non–Abelian Hodge theory described in Chapter 7, which endows the semi–simple representations of a Kähler group with such a structure.

DEFINITION 3.49. A (real) *mixed Hodge structure (MHS)* is:
- an integer lattice $H_{\mathbb{Z}}$ in a $\mathbb{R}$-vector space $H_{\mathbb{R}}$,
- an increasing weight filtration $W_\bullet$ of $H_{\mathbb{R}}$, and
- a decreasing Hodge filtration $F^\bullet$ of $H_{\mathbb{C}} = H_{\mathbb{R}} \otimes_{\mathbb{R}} \mathbb{C}$,

such that $F^\bullet$ induces a pure Hodge structure of weight $l$ on $W_l/W_{l-1}$, cf. Appendix B.

THEOREM 3.50 (Morgan [**94**], Hain [**65**]). *Let $(X,x)$ be a compact Kähler manifold with a fixed basepoint. Its basepointed 1–minimal model $\rho: M_X \to \mathcal{E}^*(X)$ admits a functorial mixed Hodge structure such that:*

*(i) The differential $d$ and the wedge product $\wedge$ in $M_X$ are morphisms of mixed Hodge structures.*

*(ii) The map $\rho: M_X \to H^*(X,\mathbb{R})$ is a morphism of mixed Hodge structures.*

Therefore, the de Rham fundamental group of a compact Kähler manifold is endowed with a functorial mixed Hodge structure depending on the base point.

The weight filtration $W_\bullet$ in the 1–minimal model is the one described in the proof of Proposition 3.20 which assigns weight $n$ to the spaces of indecomposables $V_n^1$ and which is dual to the lower central series filtration in the Malcev algebra. Conditions (i) and (ii) in Theorem 3.50 determine the mixed Hodge structure on the basepointed 1–minimal model. We illustrate this in the first two steps:

The first step is $M(1,1) \cong \wedge^*(H^1(X))$. Thus the wedges $x_1 \wedge \cdots \wedge x_k$ with $x_i \in H^1 X$ have weight $k$ assigned. For the Hodge type, as $M(1,1) \otimes \mathbb{C} \cong \wedge^* H^1(X,\mathbb{C})$, the products $x_1 \wedge \cdots \wedge x_p \wedge \cdots \wedge x_q$ with $x_1, \ldots, x_p \in H^{1,0}(X)$, $x_{p+1}, \ldots, x_q \in H^{0,1}(X)$ are of Hodge type $(p,q)$.

Now, $M(1,2) = M(1,1) \otimes \wedge(V_2^1)$, with $V_2^1 \cong \ker(H^2 M(1,1) \to H^2(X,\mathbb{R}))$. This gives $V_2^1$ its MHS, and, together with the multiplicativity of the MHS on $M_X$, the structure is determined on $M(1,2)$.

Among the consequences of Theorem 3.50 we may remark the following, which imposes restrictions on Malcev algebras of Kähler groups:

COROLLARY 3.51. *Suppose $\Gamma$ is a Kähler group. Then the linear quotients $\Gamma_n/\Gamma_{n+1} \otimes \mathbb{R}$ have even rank for odd $n$.*

REMARK 3.52. Corollary 3.51 is *not* a consequence of the fact that the Malcev algebras of Kähler groups are quadratically presented. Indeed, some of the examples of quadratically presented algebras given in [**34**] have an Abelianisation of even rank and a space of order three brackets of odd dimension.

It is also worth remarking that Theorem 3.50 is a generalisation of the Formality Theorem 3.13, because in the case of compact Kähler manifolds, the fact that the map $M_X \longrightarrow \mathcal{E}^*(X)$ respects the Hodge filtration allows the construction of a quasi–isomorphism $M_X \longrightarrow H^*(X, \mathbb{R})$, of which we have described the first stages in Proposition 3.20. In the case of a fundamental group of a compact Kähler manifold, Theorem 3.50 implies that its rational Malcev algebra admits a quadratic presentation.

Actually, Morgan [**94**] proved a stronger version of Theorem 3.50, with coefficients in $\mathbb{Q}$, which is also valid for any smooth complex algebraic variety and which shows that the full minimal model admits a (non–unique) MHS, which endows the higher homotopy groups of simply connected Kähler manifolds with mixed Hodge structures. For smooth complex algebraic manifolds, the weights in the cohomology group $H^1(X)$ are $1, 2$, and in $H^2(X)$ they are $2, 3, 4$. Therefore the images $\mathrm{Im}\,(H^2(M(1, n))) \subset H^2(X)$ stabilise after $\mathrm{Im}\,(H^2(M(1, 4)))$, and one reaches the following result, analogous to Proposition 3.20:

COROLLARY 3.53 (Morgan). *Let $X$ be a smooth complex algebraic variety. Its rational Malcev algebra $\mathcal{L}\pi_1(X)$ is a quotient of the free Lie algebra over the first homology group $L(H_1(X, \mathbb{Q}))$ by an ideal $J$ which is finitely generated by elements of bracket length $2, 3, 4$ in $L(H_1(X))$.*

Thus, for instance, the group $\Gamma$ in Example 3.23 cannot be the fundamental group of any smooth complex algebraic manifold.

CHAPTER 4

# $L^2$–cohomology of Kähler groups

## 1. Introduction

To a finitely generated group $\Gamma$ one[1] associates a numerical invariant $e(\Gamma)$, the number of ends of $\Gamma$. Roughly speaking, $e(\Gamma)$ is the number of connected components at infinity of any Cayley graph for $\Gamma$ (see subsection 2.3 below). The possible values of $e(\Gamma)$ are 0, 1, 2 and $\infty$; $e(\Gamma) = 0$ if and only if $\Gamma$ is finite, and all finite groups are Kähler, while $e(\Gamma) = 2$ if and only if $\Gamma$ is virtually infinite cyclic, and such groups are never Kähler. In this Chapter we present a proof of a theorem of Gromov [58] and of some generalisations due to Arapura–Bressler–Ramachandran [5] which imply that an infinite Kähler group has one end. This will follow from a theorem which classifies those Kähler groups $\Gamma$ for which the first reduced cohomology group (see 2.2 below) $\bar{H}^1(\Gamma, l^2(\Gamma))$ of $\Gamma$ with values in the regular representation, does not vanish.

THEOREM 4.1 ([58], [5]). *Let $\Gamma$ be a Kähler group with $\bar{H}^1(\Gamma, l^2(\Gamma)) \neq 0$. Then $\Gamma$ is commensurable with the fundamental group $\pi_1(S_g)$ of a compact orientable surface $S_g$ of genus $g \geq 2$.*

More precisely, this means that there exists a subgroup $\Gamma' \subset \Gamma$ of finite index and an exact sequence
$$1 \longrightarrow F \longrightarrow \Gamma' \longrightarrow \pi_1(S_g) \longrightarrow 1$$
with $F$ finite.

Observe that conversely any group $\Gamma$ commensurable with $\pi_1(S_g)$, $g \geq 2$, satisfies $\bar{H}^1(\Gamma, l^2(\Gamma)) \neq 0$. The proof of Theorem 4.1 is geometric and yields that if $X$ is a compact Kähler manifold with $\bar{H}^1(\Gamma, l^2(\Gamma)) \neq 0$, where $\Gamma = \pi_1(X)$, then its universal covering $\tilde{X}$ fibers properly and $\Gamma$–equivariantly over the hyperbolic disk $\mathbb{D}^2$ (see Section 2). Concerning ends, we will show in Section 4 that if $\bar{H}^1(\Gamma, l^2(\Gamma)) = 0$, then $e(\Gamma) \in \{0, 1, 2\}$. This, together with the fact that compact surface groups have one end, will imply the following positive answer to a question raised in [75]:

COROLLARY 4.2 ([5]). *A Kähler group has either 0 or 1 end; in particular it does not split as a non–trivial free product (or as a product amalgamated over a finite group).*

In Section 5 we shall prove a generalisation of Theorem 4.1. The crucial step in all the proofs is the construction of fibrations over curves from the singular holomorphic foliations defined by certain holomorphic 1–forms. This argument is a generalisation of the classical Castelnuovo–de Franchis Theorem 2.7.

---
[1] Freudenthal

We are grateful to A. Valette for discussions about $L^2$–cohomology, and to M. Ramachandran for supplying us with a proof of Stein's Theorem 4.23 and for communicating Napier's Lemma 4.32 and its proof to us.

## 2. Simplicial $L^2$–cohomology and ends

**2.1.** Let $K$ be a simplicial complex, $C^*(K)$ its complex of cochains and $C^*_{(2)}(K)$ the subspace of $l^2$–cochains, that is, the space of cochains $f$ such that

$$\sum |f(\sigma)|^2 < \infty \,,$$

where the summation is over all geometric simplices of $K$. If for every $p$ there is $C(p) < \infty$ such that every $p$–simplex is contained in at most $C(p)$ $(p+1)$–simplices, we say that $K$ has *bounded geometry*, denoted by $\mathrm{Geo}(K) < \infty$. In this case $C^*_{(2)}(K)$ is a subcomplex of $C^*(K)$ and the coboundary operators restrict to bounded operators of Hilbert spaces

$$d_p \colon C^p_{(2)}(K) \longrightarrow C^{p+1}_{(2)}(K) \,,$$

for all $p \geq 0$. In particular, the space of $l^2$–cocycles $Z^p_{(2)}(K) = \ker d_p$ is closed, whereas the subspace $B^p_{(2)}(K) = \mathrm{Im}\, d_{p-1}$ of $l^2$–coboundaries is not usually closed.

The $l^2$–cohomology of the pair $(|K|, \Gamma)$, where $\Gamma$ is a group acting simplicially on $K$, is given by the family of continuous $\Gamma$–modules

$$H^p_{(2)}(K) = Z^p_{(2)}(K)/B^p_{(2)}(K) \,,$$

for all $p \geq 0$, while its reduced $l^2$–cohomology is given by the unitary $\Gamma$–modules

$$\bar{H}^p_{(2)}(K) = Z^p_{(2)}(K)/\bar{B}^p_{(2)}(K) \,,$$

for all $p \geq 0$. Observe that $H^p_{(2)}(K)$ is endowed with a $\Gamma$–invariant Hilbert semi–norm and that $\bar{H}^p_{(2)}(K)$ is the quotient of $H^p_{(2)}(K)$ by $\overline{(0)}$.

It is a result of J. Dodziuk ([**42**] Corollary 1.3 and remark) that, if $\Gamma$ acts freely on $K$ such that $\Gamma \setminus K$ is finite, the topological $\Gamma$–module $H^*_{(2)}(K)$, and hence $\bar{H}^*_{(2)}(K)$, is a homotopy invariant of the pair $(|K|, \Gamma)$.

For a complex $K$ with $\mathrm{Geo}(K) < \infty$, the inclusion $C^*_{(2)}(K) \hookrightarrow C^*(K)$ induces a natural map $H^*_{(2)}(K) \to H^*(K)$ whose kernel, the exact part of $H^*_{(2)}(K)$, we denote $H^*_{(2),\mathrm{ex}}(K)$.

LEMMA 4.3. *The subspace $H^*_{(2),\mathrm{ex}}(K)$ is closed and the natural map $H^*_{(2)}(K) \to H^*(K)$ factors through a natural map $\bar{H}^*_{(2)}(K) \to H^*(K)$, whose kernel $\bar{H}^*_{(2),\mathrm{ex}}(K)$ is the quotient of $H^*_{(2),\mathrm{ex}}(K)$ by $\overline{(0)}$.*

PROOF. This follows immediately from the fact that evaluation on $p$–cycles defines continuous linear forms on $Z^p_{(2)}(K)$ and hence $B^p(K) \cap Z^p_{(2)}(K) \supset \overline{B^p_{(2)}(K)}$. □

**2.2.** We shall now show that if $\Gamma$ acts freely and simplicially on $K$ such that $\Gamma \setminus K$ is finite, then the $\Gamma$–module $H^1_{(2),\mathrm{ex}}(K)$ only depends on $\Gamma$. To this end we recall that given a topological vector space $E$ on which $\Gamma$ acts by continuous linear transformations, the space

$$H^1(\Gamma, E) = Z^1(\Gamma, E)/B^1(\Gamma, E)$$

is a topological vector space, as the quotient of the space of 1–cocycles $Z^1(\Gamma, E)$ endowed with the topology of pointwise convergence. The reduced first cohomology group is then
$$\bar{H}^1(\Gamma, E) = Z^1(\Gamma, E)/\overline{B^1(\Gamma, E)} \ .$$
Let $\rho$ denote the right action of $\Gamma$ on $l^2(\Gamma)$ and $H^1(\Gamma, l^2(\Gamma))$ the corresponding first cohomology. The left action of $\Gamma$ on $l^2(\Gamma)$ induces then an action of $\Gamma$ on $H^1(\Gamma, l^2(\Gamma))$.

PROPOSITION 4.4 ([11]). *Let $K$ be a connected complex and $\Gamma \times K \to K$ a free simplicial action such that $\Gamma \setminus K$ is finite. There are $\Gamma$–isomorphisms of topological vector spaces:*
$$\begin{aligned} H^1(\Gamma, l^2(\Gamma)) &\simeq H^1_{(2),\mathrm{ex}}(K) \\ \bar{H}^1(\Gamma, l^2(\Gamma)) &\simeq \bar{H}^1_{(2),\mathrm{ex}}(K) \ . \end{aligned}$$

PROOF. According to Lemma 4.3, the second isomorphism follows from the first one.

The $\Gamma$–space $H^1_{(2),\mathrm{ex}}(K)$ only depends on the 1–skeleton $K^{(1)}$ of $K$, and by taking a maximal tree in the graph $\Gamma \setminus K^{(1)}$ we may $\Gamma$–equivariantly retract $K^{(1)}$ to a Cayley graph $C$ of $\Gamma$, thus obtaining an isomorphism $H^1_{(2),\mathrm{ex}}(K) \simeq H^1_{(2),\mathrm{ex}}(C)$.

Let $S = S^{-1}$ be a finite generating set of $\Gamma$ defining $C$, and let $\mathbb{C}(\Gamma)$ be the space of all functions on $\Gamma$. Since $H^1(\Gamma, \mathbb{C}(\Gamma)) = 0$, we may for every $c \in Z^1(\Gamma, l^2(\Gamma))$ choose $f \in \mathbb{C}(\Gamma) = C^0(C)$ such that
$$c(\gamma) = \rho(\gamma)f - f \ , \ \forall \, \gamma \in \Gamma \ ,$$
and define
$$\alpha_c = \sum_{s \in S} c(s) = df \in C^1_{(2)}(C) \ .$$
It is then an easy verification that the map $c \to \alpha_c$ induces a $\Gamma$–isomorphism of topological vector spaces $H^1(\Gamma, l^2(\Gamma)) \simeq H^1_{(2),\mathrm{ex}}(C)$. □

**2.3.** The space of ends $E(X)$ of a locally compact separable metric space $X$ is the inverse limit
$$\lim_{\substack{K \subset X \\ \mathrm{compact}}} \pi_0(X \setminus K)$$
and, as such, $E(X)$ is a totally disconnected topological space. When $X$ is connected and locally connected, the space $E(X)$ is compact. For a group $\Gamma$ acting freely on a connected simplicial complex $K$ such that $\Gamma \setminus K$ is finite, the homeomorphism type of $E(K)$ only depends on $\Gamma$, and we denote by $e(\Gamma)$ its cardinality. The following basic facts are due to Hopf and Freudenthal:
1. $e(\Gamma) \in \{0, 1, 2, \infty\}$.
2. $e(\Gamma) = 0$ if and only if $\Gamma$ is finite.
3. $e(\Gamma) = 2$ if and only if $\Gamma$ contains an infinite cyclic subgroup of finite index.

EXAMPLE 4.5. (1) $\mathbb{Z}$ has two ends.
(2) The free group on two generators $F_2$ has infinitely many ends.
(3) The modular group
$$SL(2, \mathbb{Z}) \cong \mathbb{Z}_4 \underset{\mathbb{Z}_2}{*} \mathbb{Z}_6 = \langle s, t | s^4 = 1 = t^6, s^2 = t^3 \rangle$$
has infinitely many ends.
(4) The Heisenberg group $\mathcal{H}_3(\mathbb{Z})$ has one end.

(5) Surface groups $\Gamma_g = \langle a_1, \ldots, b_g | [a_1, b_1] \ldots [a_g, b_g] \rangle$ have one end each.

Our next task is to establish the inequality $e(\Gamma) - 1 \leq \dim H^1(\Gamma, l^2(\Gamma))$ for all finitely generated groups $\Gamma$. Let $X$ be any complex, $H_f^*(X)$ its cohomology with finite support, and $H_{f,\mathrm{ex}}^*(X)$ the kernel of the natural map $H_f^*(X) \to H^*(X)$.

PROPOSITION 4.6 (compare with [11]). *Suppose that $X$ is a connected graph with $\mathrm{Geo}(X) < \infty$, and let $\mathcal{S}(E(X))$ be the space of locally constant functions on $E(X)$.*

1. *There is a natural isomorphism $\mathcal{S}(E(X))/\mathbb{C} \to H^1_{f,\mathrm{ex}}(X)$.*
2. *The natural map $H^1_f(X) \to H^1_{(2)}(X)$ is injective.*
   *Both maps are equivariant with respect to simplicial automorphisms of $X$.*
3. *Let $\Gamma$ be a finitely generated group. Then $e(\Gamma) - 1 \leq \dim H^1(\Gamma, l^2(\Gamma))$.*

PROOF. 1. For $f \in \mathcal{S}(E(X))$ let $\{a_1, \ldots, a_n\}$ be the finite set of values of $f$ and $E_i = f^{-1}(a_i)$. Choose a finite set of vertices $F \subset X^0$ such that if $X_F$ denotes the subgraph of $X$ generated by $X^0 \setminus F$, each $E_i$ is the set of ends of a subgraph $X_F^i \subset X_F$ which is a union of infinite connected components of $X_F$. Define $h = \sum_{i=1}^n a_i \chi_{W_i}$, where $\chi_{W_i}$ is the characteristic function of the vertex set $W_i$ of $X_F^i$. Observe that a different choice of $F \subset X^0$ leads to a function $h'$ with $h - h' \in C_f^0(X)$. Therefore the cohomology class of the cocycle

$$dh \in Z_f^1(X)$$

only depends on $f$. The equality $dh = dg$ for some $g \in C_f^0(X)$ implies that $h - g$ is constant, say $h - g = \lambda \cdot 1_{X^0}$, and hence $a_i = \lambda$ for all $1 \leq i \leq n$ since the support of $g$ is finite and all the $W_i$ are infinite. This shows that $\mathcal{S}(E(X))/\mathbb{C}$ injects into $H^1_{f,\mathrm{ex}}(X)$. Surjectivity is left as an exercise for the reader.
2. Let $c \in Z_f^1(X)$ with $c = df$ and $f \in C_{(2)}^0(X)$. Then $f$ is constant on all connected components of the subgraph $X'$ generated by $X^{(1)} \setminus \mathrm{supp}\,(c)$ and hence $f$ is zero on each such infinite component. This implies $f \in C_f^0(X)$ and proves 2.
3. Apply Proposition 4.4 and 1. and 2. above to any Cayley graph $X$ of $\Gamma$. $\square$

**2.4.** In this subsection we show that if $\Gamma$ is a finitely generated group with infinitely many ends, then $H^1(\Gamma, l^2(\Gamma)) = \bar{H}^1(\Gamma, l^2(\Gamma))$. This may be deduced from Stallings's structure theorem [126] together with Guichardet's characterisation of non–amenable infinite groups [63]. An argument not using Stallings's theorem is due to Soardi and Woess [123]. We present here a direct argument due to Ch. Pittet. We say that a locally finite graph $X$ satisfies a *linear isoperimetric inequality* (l. i. i.) if there exists an $\varepsilon > 0$ such that

$$\frac{|\partial F|}{|F|} \geq \varepsilon$$

for any finite set $F \subset X^0$ of vertices. Here $\partial F$ denotes the set of vertices in $F$ connected to some vertex in $X^0 \setminus F$. We recall the following classical fact:

Let $X$ be a connected graph with $\mathrm{Geo}(X) < \infty$ and assume that $X$ satisfies an l. i. i. Then the image of $d \colon C_{(2)}^0(X) \to C_{(2)}^1(X)$ is closed. Equivalently, $H^1_{(2)}(X) = \bar{H}^1_{(2)}(X)$.

THEOREM 4.7 (Pittet). *Let $X$ be a connected locally finite graph with $\mathrm{Geo}(X) < \infty$. Assume that there exists an $r > 0$ such that, for every $x \in X^0$, the subgraph* generated by $X^0 \setminus B(x,r)$ has at least three unbounded connected components. Then $X$ satisfies a linear isoperimetric inequality, in particular $H^1_{(2)}(X) = \bar{H}^1_{(2)}(X)$.

SKETCH OF PROOF. Details can be found in [**101**]. The proof consists of the following observations:

(a) Let $Y$ be a graph and $Z \colon Y^0 \to Y^0$ a map such that $d(Z(v), v) \leq 1$ and $|Z^{-1}(v)| \geq 2$, for all $v \in Y^0$. Then $Y$ satisfies an l. i. i.

Indeed, for $\Omega \subset Y^0$ a finite set and Int $\Omega = \Omega \setminus \partial\Omega$ we have $Z^{-1}(\text{Int } \Omega) \subset \Omega$ and hence $2|\text{Int } \Omega| \leq |Z^{-1}(\text{Int } \Omega)| \leq |\Omega|$, which implies

$$\frac{|\partial \Omega|}{|\Omega|} \geq \frac{1}{2}.$$

(b) The property of admitting an l. i. i. is invariant under quasi–isometries.

(c) Let $X$ and $r$ be as in the statement of the Theorem. Fix $m > 4r + 2$ and take $V \subset X^0$ maximal such that $d(u, v) \geq m$ for all $u \neq v \in V$. Fix $v_0 \in V$ and define $Z \colon V \to V$ as follows:

- for $v \in V$ with $d(v_0, v) \leq m$, set $Z(v) = v_0$; and
- for $v \in V$ with $d(v_0, v) \geq m + 1$, fix a geodesic $g_v$ in $X$ connecting $v$ to $v_0$, take $x \in g_v$ at distance $m + 1$ from $v$ and let $Z(v) \in V$ be such that $d(Z(v), x) = d(V, x)$.

One shows then that $d(Z(v), v) \leq 2m + 1$ and $|Z^{-1}(v)| \geq 2 \ \forall v \in V$.

Put on $V$ the graph structure $\mathcal{V}$ such that $u, v \in V$ are adjacent if and only if $d(u, v) \leq 2m + 1$. The graph $\mathcal{V}$ satisfies an l. i. i. by (a) and (c). Since $\mathcal{V}$ is quasi–isometric to $X$, (b) implies that $X$ satisfies an l. i. i. □

COROLLARY 4.8. *Let $\Gamma$ be a finitely generated group with infinitely many ends. Then $\bar{H}^1(\Gamma, l^2(\Gamma)) = H^1(\Gamma, l^2(\Gamma))$ and is infinite–dimensional.*

PROOF. Let $X$ be a Cayley graph of $\Gamma$ with respect to a finite generating set. Since $e(\Gamma) = \infty$, the graph $X$ satisfies the hypothesis of Theorem 4.7, in particular $H^1_{(2)}(X) = \bar{H}^1_{(2)}(X)$, and hence $H^1_{(2),\text{ex}}(X) = \bar{H}^1_{(2),\text{ex}}(X)$, which implies $H^1(\Gamma, l^2(\Gamma)) = \bar{H}^1(\Gamma, l^2(\Gamma))$ by Proposition 4.4. Since $e(\Gamma) - 1 \leq \dim H^1(\Gamma, l^2(\Gamma))$ by Proposition 4.6 3., we conclude that $\bar{H}^1(\Gamma, l^2(\Gamma))$ is infinite–dimensional. □

## 3. de Rham $L^2$–cohomology

Let $(X, g)$ be an oriented complete Riemannian manifold and $*$ its Hodge operator. The space $\Omega^p_{(2)}(X)$ of measurable $p$–forms $\alpha$ satisfying

$$\int_X \alpha \wedge *\alpha < \infty$$

is a Hilbert space on which $\text{Iso}(X)$, the group of isometries of $X$, acts unitarily. Considering $\Omega^p_{(2)}(X)$ as a subspace of the space of $p$–currents we define

$$\begin{aligned} Z^p_{(2)}(X) &= \left\{\omega \in \Omega^p_{(2)}(X) | d\omega = 0\right\} \\ B^p_{(2)}(X) &= d\left(\Omega^{p-1}_{(2)}(X)\right) \cap \Omega^p_{(2)}(X) \,.\end{aligned}$$

Then $Z^p_{(2)}(X)$ is a closed subspace of the Hilbert space $\Omega^p_{(2)}(X)$ and hence contains $\overline{B^p_{(2)}(X)}$. The $L^2$–cohomology groups of the Riemannian manifold $(X, g)$ are given

by
$$H^p_{(2)}(X) = Z^p_{(2)}(X)/B^p_{(2)}(X) ,$$
for all $p \geq 0$, while its reduced $L^2$–cohomology groups are
$$\bar{H}^p_{(2)}(X) = Z^p_{(2)}(X)/\overline{B^p_{(2)}(X)} ,$$
for all $p \geq 0$. Both $H^*_{(2)}(X)$ and $\bar{H}^*_{(2)}(X)$ are continuous $\text{Iso}(X)$–modules and $\text{Iso}(X)$ acts unitarily on the latter.

The space $\bar{H}^p_{(2)}(X)$ admits the following description via harmonic forms. Let $\triangle = d^*d + dd^*$ be the Laplacian acting on currents,
$$\mathcal{H}^p_{(2)}(X) = \{\omega \in \Omega^p_{(2)}(X) | \triangle\omega = 0\}$$
and
$$\mathcal{B}^p_{(2)}(X) = d^*(\Omega^{p+1}_{(2)}(X)) \cap \Omega^p_{(2)}(X) .$$

THEOREM 4.9 (Hodge decomposition [107]). *Let $(X, g)$ be an oriented complete Riemannian manifold. The following orthogonal sum decompositions hold:*
$$\Omega^p_{(2)}(X) = \mathcal{H}^p_{(2)}(X) \oplus \overline{B^p_{(2)}(X)} \oplus \overline{\mathcal{B}^p_{(2)}(X)}$$
$$Z^p_{(2)}(X) = \mathcal{H}^p_{(2)}(X) \oplus \overline{B^p_{(2)}(X)} .$$

The Hodge decomposition implies that the $\text{Iso}(X)$–action on $\bar{H}^p_{(2)}(X)$ is unitarily equivalent to the $\text{Iso}(X)$–action on $\mathcal{H}^p_{(2)}(X)$.

We turn now to Dodziuk's version of de Rham's theorem. Let $\Gamma$ be a group acting freely, properly discontinuously and by isometries on $X$ such that $\Gamma \backslash X$ is compact; let $K$ be a triangulation of $X$ which is a lift of a finite triangulation of $\Gamma \backslash X$.

THEOREM 4.10 ([42], Theorem 1). *Integration of forms over simplices of $K$ induces a $\Gamma$–isomorphism of Hilbert spaces*
$$\mathcal{H}^p_{(2)}(X) \xrightarrow{\cong} \bar{H}^p_{(2)}(K) .$$

Notice that, for forms in $\mathcal{H}^p_{(2)}(X)$, integration over simplices is well–defined since such forms are $C^\infty$ by Weyl's lemma.

Dodziuk's theorem has the following immediate consequence:

COROLLARY 4.11. *Let $\mathcal{H}^1_{(2),\text{ex}}(X)$ denote the space of exact, square–integrable harmonic 1-forms. There is a $\Gamma$–isomorphism*
$$\bar{H}^1(\Gamma, l^2(\Gamma)) \cong \mathcal{H}^1_{(2),\text{ex}}(X) .$$
*In particular, if $\Gamma$ has infinitely many ends, $\dim \mathcal{H}^1_{(2),\text{ex}}(X) = \infty$.*

PROOF. Let $K$ be a triangulation of $X$ as in Theorem 4.10. De Rham's theorem and Theorem 4.10 imply that the spaces $\bar{H}^1_{(2),\text{ex}}(K)$ and $\mathcal{H}^1_{(2),\text{ex}}(X)$ are $\Gamma$–isomorphic. The Corollary follows then from Proposition 4.4 and from Corollary 4.8. □

For later use we record

LEMMA 4.12. *Let $(X, g)$ be a connected, oriented complete Riemannian manifold. Then*
$$\mathcal{H}^1_{(2),\text{ex}}(X) = \{\alpha \in \Omega^1_{(2)}(X) | \alpha = df \text{ and } f \text{ is harmonic}\} .$$

PROOF. For $\alpha \in \mathcal{H}^1_{(2),\text{ex}}(X)$, $\alpha = df$, we have $d\triangle f = 0$ and hence $\triangle f$ is a constant, which we denote by $c$. Fix some ball $B \subset X$ and choose a sequence $(\varphi_n)_{n \in \mathbb{N}}$ of $C^\infty$-functions on $X$ with compact support such that $0 \leq \varphi_n \leq 1$, $\varphi_n|_B = 1$, and $|d_x\varphi_n|^2 \leq \frac{1}{n}\varphi_n(x)$, for all $x \in X$. Then we have

$$|c|^2 \int_X \varphi_n dv(x) = \langle \triangle f, \varphi_n \triangle f \rangle = \langle df, (\triangle f) \cdot d\varphi_n \rangle$$

$$\leq |c| \, \|d\varphi_n\|_2 \cdot \|df\|_2$$

$$\leq |c| \frac{1}{\sqrt{n}} \left( \int_X \varphi_n dv(x) \right)^{1/2} \|df\|_2$$

and hence $|c|(\int \varphi_n)^{1/2} \leq \frac{1}{\sqrt{n}}\|df\|_2$, for all $n \in \mathbb{N}$, which implies $c = 0$ since $\int \varphi_n \geq \text{Vol}(B) > 0$. $\square$

## 4. Fibering Kähler manifolds over $\mathbb{D}^2$

Let $(X, g)$ be a complex manifold of dimension $n$ with a Hermitian metric $g$.

DEFINITION 4.13. $(X, g)$ has *bounded geometry* (denoted by $\text{Geo}(X) < \infty$) if there exist constants $C_1, C_2, R > 0$ such that every point $x \in X$ has a neighbourhood $U_x$ admitting a biholomorphism

$$\psi_x \colon U_x \longrightarrow B(0, R) \subset \mathbb{C}^n$$

such that

$$C_1 \psi_x^* g_E \leq g_x \leq C_2 \psi_x^* g_E \; ,$$

where $g_E$ is the standard metric on $\mathbb{C}^n$.

THEOREM 4.14 ([58], [5]). *Let $X$ be a complete Kähler manifold with bounded geometry. Assume that $H^1(X, \mathbb{R}) = 0$ and $\mathcal{H}^1_{(2)}(X) \neq 0$. Then there exists a proper holomorphic map to the hyperbolic disk $\mathbb{D}^2$*

$$h \colon X \longrightarrow \mathbb{D}^2$$

*with connected fibers. Moreover*

1. *the fibers of $h$ are permuted by $\text{Aut } X$, and*
2. *the map $h$ induces an isomorphism*

$$h^* \colon \mathcal{H}^1_{(2)}(\mathbb{D}^2) \longrightarrow \mathcal{H}^1_{(2)}(X) \; .$$

Before going into the proof, we show how to deduce Theorem 4.1 from Theorem 4.14 and the results in Sections 2 and 3.

PROOF OF THEOREM 4.1. Let $X$ be a compact Kähler manifold with $\Gamma = \pi_1(X)$, and $\widetilde{X}$ its universal covering. Since $\bar{H}^1(\Gamma, l^2(\Gamma)) \neq 0$, we have $\mathcal{H}^1_{(2)}(\widetilde{X}) \neq 0$ by Corollary 4.11. Let $h \colon \widetilde{X} \to \mathbb{D}^2$ be the holomorphic map obtained by applying Theorem 4.14 to $\widetilde{X}$. It follows then from 1. in Theorem 4.14 that $\Gamma$ acts properly discontinuously on $\mathbb{D}^2$ with $\Gamma \setminus \mathbb{D}^2$ compact, and hence $\Gamma$ is commensurable with the fundamental group of a compact surface of genus $g \geq 2$. $\square$

REMARK 4.15. (cf. Gromov [**61**], 7.D) A Kähler group which is quasi–isometric to a compact surface group of genus $\geq 2$ is commensurable with such a group. This follows from Theorem 4.1 and the quasi–isometry invariance of of the non–vanishing of $L^2$ Betti numbers.

The rest of this Section is devoted to the proof of Theorem 4.14 and is independent of Sections 2 and 3.

**4.1. Singular holomorphic foliations.** Let $X$ be a complete Kähler manifold. We have the Hodge decomposition

$$\Omega^p_{(2)}(X) = \mathcal{H}^p_{(2)}(X) \oplus \overline{\mathcal{B}^p_{(2)}(X)} \oplus \overline{\mathcal{B}^p_{(2)}(X)}$$

and the decomposition into Hodge types

$$\Omega^p_{(2)}(X) = \bigoplus_{r+s=p} \Omega^{(r,s)}_{(2)}(X) \ .$$

Since the Laplacian preserves Hodge types, we obtain a decomposition

$$\mathcal{H}^1_{(2)}(X) = \mathcal{H}^1_{(2)}(X) \cap \Omega^{(1,0)}_{(2)}(X) \oplus \mathcal{H}^1_{(2)}(X) \cap \Omega^{(0,1)}_{(2)}(X) \ .$$

Moreover, $\mathcal{H}^1_{(2)}(X) = \{\alpha \in \Omega^1_{(2)}(X) | d\alpha = 0, d^*\alpha = 0\}$ by Gaffney's version of Stokes's theorem, and we therefore deduce that $\mathcal{H}^1_{(2)}(X) \cap \Omega^{(1,0)}_{(2)}(X)$ consists of holomorphic 1–forms.

LEMMA 4.16 ([**59**]). *Let $X$ be a complete Kähler manifold such that $\mathcal{H}^*_{(2)}(X) \subset \mathcal{H}^*_{(\infty)}(X)$. Let $\alpha_1$ and $\alpha_2$ be real 1–forms on $X$ such that $\alpha_i \in \mathcal{H}^1_{(2),\mathrm{ex}}(X)$ and let $\omega_i$ be the $(1,0)$–component of $\alpha_i$. Then $\omega_i$ is closed, holomorphic and square–integrable, and*

$$\omega_1 \wedge \omega_2 \equiv 0.$$

PROOF. The fact that $\omega_i$ is closed, holomorphic and square–integrable follows from the remarks preceeding the Lemma. The proof of $\omega_1 \wedge \omega_2 \equiv 0$ proceeds in two steps.

(i) Let $f\colon X \to \mathbb{R}$ be a smooth function with $df \in \mathcal{H}^1_{(2)}(X)$, and let $\psi \in \mathcal{H}^i_{(2)}(X)$. Then we claim that $df \wedge \psi \in \overline{B^{i+1}_{(2)}(X)}$.

Since $\mathcal{H}^i_{(2)}(X) \subset \mathcal{H}^i_{(\infty)}(X)$, $\psi$ is bounded, and hence $df \wedge \psi$ is square–integrable. For $c > 0$, let $f_c\colon X \to \mathbb{R}$ be the unique continuous function which coincides with $f$ on $V_c = \{x \in X \colon |f(x)| < c\}$ and is constant otherwise. Stokes's theorem implies that $df_c = \chi_{V_c} \cdot df$, where $\chi_{V_c}$ denotes the characteristic function of $V_c$. Setting $\beta_c = f_c \wedge \psi \in \Omega^i_{(2)}(X)$, we have $d\beta_c = \chi_{V_c}(df \wedge \psi)$, and hence $\lim_{c\to\infty} d\beta_c = df \wedge \psi$ in the $L^2$–norm. This shows that $df \wedge \psi \in \overline{B^{i+1}_{(2)}(X)}$.

(ii) Let $\alpha_j = df_j$, $f_j\colon X \to \mathbb{R}$, and $\omega_j$ as in the statement of the Lemma; and let

$$\omega_j = df_j + i\beta_j$$

be the decomposition of $\omega_j$ into real and imaginary parts. We have

$$\omega_1 \wedge \omega_2 = (df_1 \wedge df_2 - \beta_1 \wedge \beta_2) + i(df_1 \wedge \beta_2 + \beta_1 \wedge df_2).$$

Since $\omega_j$ is holomorphic and square integrable, we have $\beta_j \in \mathcal{H}^1_{(2)}(X)$. Hence by (i)

$$\mathrm{Im}\,(\omega_1 \wedge \omega_2) = df_1 \wedge \beta_2 + \beta_1 \wedge df_2 \in \overline{B^2_{(2)}(X)} \ .$$

On the other hand, as $\omega_1 \wedge \omega_2$ is holomorphic, Im $(\omega_1 \wedge \omega_2)$ is harmonic and therefore belongs to $\mathcal{H}^2_{(2)}(X)$. The Hodge decomposition implies then Im $(\omega_1 \wedge \omega_2) \equiv 0$ and hence $\omega_1 \wedge \omega_2 \equiv 0$. □

Let $X$ be a complex manifold and $\omega$ a closed holomorphic 1–form. For every open set $U \subset X$ on which $\omega$ is exact, let $\omega|_U = df_U$ with $f_U: U \to \mathbb{C}$ holomorphic, and define $\mathcal{R}_U$ to be the equivalence relation on $U$ given by the level sets of $f_U$. Here and in the sequel, the level sets of a map are the connected components of its fibers. The equivalence relation $\mathcal{R}_U$ only depends on $\omega|_U$, and we let $\mathcal{R}_\omega$ denote the equivalence relation on $X$ generated by the equivalence relations $\mathcal{R}_U$. Observe that outside the vanishing locus of $\omega$, the leaves of $\mathcal{R}_\omega$ are immersed connected complex submanifolds.

LEMMA 4.17. *Let $\omega_1$ and $\omega_2$ be closed holomorphic 1–forms such that $\omega_i \not\equiv 0$ and $\omega_1 \wedge \omega_2 \equiv 0$. Then $\mathcal{R}_{\omega_1} = \mathcal{R}_{\omega_2}$.*

PROOF. This being a local question, we may assume $\omega_i = df_i$ where $f_i: U \to \mathbb{C}$ is a holomorphic function defined on an open connected subset $U \subset \mathbb{C}^n$. Observe that, by Sard's theorem, $f_1$ and $f_2$ are locally constant on $A = \{x \in U | d_x f_1 - d_x f_2 = 0\}$. Since $\omega_1 \wedge \omega_2 \equiv 0$, the map

$$F: U \setminus A \longrightarrow \mathbb{C}^2$$
$$x \longmapsto (f_1(x), f_2(x))$$

has constant rank $= 1$.

Let $C \subset U \setminus A$ be an infinite connected set on which $f_1$ is constant with value $c_1 \in \mathbb{C}$, let $x_0 \in C$ and $U_{x_0}$ be a neighbourhood of $x_0$ in $U \setminus A$ such that $F(U_{x_0})$ is a 1–dimensional submanifold of $\mathbb{C}^2$. We have:

$$F(U_{x_0} \cap C) \subset (\{c_1\} \times \mathbb{C}) \cap F(U_{x_0}).$$

If $f_2|_C$ is not constant in a neighbourhood of $x_0$ in $C$, then $\{c_1\} \times \mathbb{C}$ and $F(U_{x_0})$ have infinitely many intersection points in some neighbourhood of $F(x_0)$ and hence coincide in a neighbourhood of $F(x_0)$. But this implies that $f_1$ is constant in a neighbourhood of $x_0$, contradicting the assumption that $df_1$ is not identically 0. Hence $f_2|_C$ is locally constant and hence constant.

Let $N$ be a level set of $f_1: U \to \mathbb{C}$. By the preceeding discussion, $f_2|_{N \setminus A}$ and $f_2|_{N \cap A}$ are locally constant, and $N$, being an analytic set, is arc–connected. This implies that $f_2|_N$ is constant. □

In the spirit of Metatheorem 1.2, we have:

SCHOLIUM 4.18. *Let $X$ be a complete Kähler manifold such that $\mathcal{H}^*_{(2)}(X) \subset \mathcal{H}^*_{(\infty)}(X)$. Assume $\mathcal{H}^1_{(2),\mathrm{ex}}(X) \neq 0$. Then $X$ admits a canonical singular holomorphic codimension 1 foliation whose leaves are permuted by $\mathrm{Aut}\, X$.*

Under certain assumptions one can show that all leaves of such a foliation are compact. We describe a tool for this in the following subsections 4.2 and 4.3.

### 4.2. Volumes.

PROPOSITION 4.19. *Let $X$ be a complete Kähler manifold with bounded geometry.*

1. *Any $p$–dimensional analytic submanifold $A \subset X$ with $\mathrm{Vol}\,(A) < \infty$ is compact.*

2. Given $v > 0$ there exists an $R(v) > 0$ such that for any connected submanifold $A \subset X$ with $\text{Vol}(A) \leq v$ and any $x \in A$, $A \subset \bar{B}(x, R(v))$.

LEMMA 4.20. *Let $A \subset \mathbb{C}^n$ be a $p$-dimensional complex submanifold containing $0$. Then, for all $R \geq 0$,*

$$\text{Vol}(A \cap B(0, R)) \geq \text{Vol}(T_0 A \cap B(0, R)) = c(p) R^{2p},$$

*where $T_0 A$ is the tangent space of $A$ at $0$ and $c(p) > 0$ only depends on $p$.*

PROOF OF LEMMA. This is a computation applying Stokes's Theorem. Let $\omega = \frac{1}{4} dd^c |\zeta|^2$, where $|\zeta|^2 = z_1 \bar{z}_1 + \cdots + z_n \bar{z}_n$, be the standard Kähler form on $\mathbb{C}^n$, and consider the form $\omega_0 = \frac{1}{4} dd^c(\log |\zeta|^2)$ defined on $\mathbb{C}^n \setminus \{0\}$. The form $\omega_0$ is the pullback of the Kähler form given by the Fubini–Study metric on $\mathbb{C}P^{n-1}$, thus it is semi–positive definite.

The forms $\omega$ and $\omega_0$ are related outside the origin by the identity

$$\omega_0 = \frac{\omega}{|\zeta|^2} - \frac{d|\zeta|^2 \wedge d^c|\zeta|^2}{|\zeta|^4}.$$

Set $A_t = B(0, t) \cap A$, for all $t \geq 0$. Since $\omega^p = d(\frac{1}{4} d^c |\zeta|^2 \wedge \omega^{p-1})$, Stokes's theorem implies that for $R \geq r > 0$:

$$\frac{1}{R^{2p}} \int_{A_R} \omega^p - \frac{1}{r^{2p}} \int_{A_r} \omega^p = \int_{\partial(A_R \setminus A_r)} \frac{1}{4} \frac{d^c|\zeta|^2 \wedge \omega^{p-1}}{|\zeta|^{2p}} =$$

$$\int_{\partial(A_R \setminus A_r)} \frac{1}{4} \frac{d^c|\zeta|^2}{|\zeta|^2} \wedge \left( \frac{\omega}{|\zeta|^2} \right)^{p-1} = \int_{\partial(A_R \setminus A_r)} \frac{1}{4} d^c \log |\zeta|^2 \wedge \omega_0^{p-1} = \int_{A_R \setminus A_r} \omega_0^p \geq 0.$$

In conclusion,

$$\frac{1}{R^{2p}} \int_{A_R} \omega^p \geq \frac{1}{r^{2p}} \int_{A_r} \omega^p$$

for $R \geq r > 0$. Moreover, $\frac{1}{r^{2p}} \int_{A_r} \omega^p = \int_{\frac{1}{r} A_r} \omega^p$, and the latter expression converges to $\text{Vol}(T_0 A \cap B(0, 1))$ as $r \to 0$, since $\frac{1}{r} A_r \to T_0 A \cap B(0, 1)$ in the space of currents. □

REMARK 4.21. The previous Lemma holds for analytic spaces, see [35], 15.1.

PROOF OF PROPOSITION 4.19. It follows from Lemma 4.20 that there exists an $r > 0$ such that for any $\varepsilon, 0 < \varepsilon < r$, there exists a $c(\varepsilon) > 0$ such that

$$\text{Vol}(B(x, \varepsilon) \cap A) \geq c(\varepsilon), \, \forall \, x \in A.$$

Hence, if $x_1, \ldots, x_n \in A$ such that the sets $B(x_i, \varepsilon) \cap A$, $1 \leq i \leq n$ are pairwise disjoint, then $\text{Vol}(A) \geq n \cdot c(\varepsilon)$, which implies that $A$ is compact and proves 1.

Fix $\varepsilon > 0$, and assume that $A$ is connected with $\text{Vol}(A) \leq v$. Choose $n$ maximal such that there are $n$ points $x_1, \ldots, x_n \in A$ for which the sets $B(x_i, \varepsilon) \cap A$, $1 \leq i \leq n$, are pairwise disjoint. Then $\bigcup_{i=1}^{n} \bar{B}(x_i, 2\varepsilon) \supset A$; since $A$ is connected, this implies that $d(x_i, x_j) \leq 4\varepsilon n \leq \frac{4\varepsilon v}{c(\varepsilon)}$, for all $i, j$ and hence $A \subset \bar{B}(x, \frac{4\varepsilon v}{c(\varepsilon)} + 4\varepsilon)$, $\forall x \in A$, which proves 2. □

## 4.3. Convergence of leaves.
In this subsection we establish a convergence result about sequences of analytic sets in a complex manifold.

Let $X$ be a metrisable complete space and $\mathcal{C}(X)$ the space of closed subsets of $X$. Given a sequence $(L_n)_{n\in\mathbb{N}}$ in $\mathcal{C}(X)$, define

$$\liminf(L_\nu) = \{x \in X |\text{ every neighbourhood } U \text{ of } x \text{ meets all but finitely many } L_\nu\}$$
$$\limsup(L_\nu) = \{x \in X |\text{ every neighbourhood } U \text{ of } x \text{ meets infinitely many } L_\nu\} \ .$$

Then $\liminf(L_\nu) \subseteq \limsup(L_\nu)$ and $\liminf(L_\nu)$ is a closed subset of $X$. We say that the sequence $(L_n)_{n\in\mathbb{N}}$ converges to $L \in \mathcal{C}(X)$, if

$$L = \liminf(L_\nu) = \limsup(L_\nu).$$

According to [135], the above notion of convergence is given by a topology on $\mathcal{C}(X)$ for which $\mathcal{C}(X)$ is compact metrisable.

Now let $X$ be a complex manifold and let $A$ and $B$ be pure–dimensional analytic subsets of $X$.

DEFINITION 4.22. $A$ and $B$ are in *relative general position* if the dimension of every irreducible component of $A \cap B$ is $\dim A + \dim B - \dim X$.

THEOREM 4.23 (Stein [129]). *Let $X$ be a complex manifold and $L$ and $L_n$ ($n \in \mathbb{N}$) be pairwise distinct pure $p$-dimensional connected analytic subsets of $X$. Assume that*

1. *any two intersecting irreducible components of $L$ are in relative general position,*
2. $\bigcup_{n=1}^{\infty} L_n$ *is an analytic subset of $X \setminus L$, and*
3. $\liminf L_n \neq \emptyset$.

*Then a subsequence of $(L_n)_{n\in\mathbb{N}}$ converges to $L$.*

Before turning to the proof of Theorem 4.23, we verify that level sets of holomorphic functions satisfy the hypothesis 1. This will be used in the proof of Gromov's Theorem 4.14.

LEMMA 4.24. *Let $\Omega \subset \mathbb{C}^n$ be an open set, $f \colon \Omega \to \mathbb{C}$ a holomorphic function, $L$ a level set of $f$ and $L_1$ and $L_2$ distinct irreducible components of $L$ with $L_1 \cap L_2 \neq \emptyset$. Then $L_1 \cap L_2$ is of pure dimension $n-2$.*

PROOF. We may assume that $0 \in L_1 \cap L_2$ and that $0$ is isolated in $(L_1 \cup L_2) \cap \{(z', z_n) \in \mathbb{C}^{n-1} \times \mathbb{C} \,|\, z' = 0\}$. Let $F_1(z', z_n)$ and $F_2(z', z_n)$ be Weierstrass polynomials describing, respectively, $L_1$ and $L_2$ in a neighbourhood of $0$. Their resultant $z' \to R(F_1(z', \cdot), F_2(z', \cdot))$ is holomorphic in a neighbourhood of $0$ in $\mathbb{C}^{n-1}$ and vanishes at $0$. Moreover, since $L_1$ and $L_2$ are distinct, this resultant does not vanish identically, and hence its zero set is pure $(n-2)$-dimensional. Hence $L_1 \cap L_2$ is $(n-2)$-dimensional at $0$. □

LEMMA 4.25. *Let $\Omega \subset \mathbb{C}^n$ be an open neighbourhood of $0$, and $A$, $(A_i)_{i\in\mathbb{N}}$ be pure $p$-dimensional analytic subsets of $\Omega$ with $\lim_{i\to\infty} A_i = A$ in $\mathcal{C}(\Omega)$. Assume that $0 \in A$ and $0$ is isolated in $A \cap (0') \times \mathbb{C}^{n-p}$. Then, for $i$ large enough we have*

$$A_i \cap (0') \times \mathbb{C}^{n-p} \neq \emptyset \ .$$

PROOF. Since 0 is isolated in $A \cap (0') \times \mathbb{C}^{n-p}$ there exist $r', r'' > 0$ such that $\overline{B'}(0, r') \times \overline{B''}(0, r'') \subset \Omega$ and $(\overline{B'}(0, r') \times S''(0, r'')) \cap A = \emptyset$, where $B', B''$ refers to open balls in $\mathbb{C}^p$, respectively $\mathbb{C}^{n-p}$. Since $\lim A_j = A$, there exists $j_0$ such that for all $j \geq j_0$: $(B'(0, r') \times B''(0, r'')) \cap A_j \neq \emptyset$ and $(\overline{B'}(0, r') \times S''(0, r'')) \cap A_j = \emptyset$. For the projection $\pi : \mathbb{C}^n \to \mathbb{C}^p$ we have that

$$\pi|_{A_j} : A_j \cap (B'(0, r') \times B''(0, r'')) \longrightarrow B'(0, r')$$

is proper for all $j \geq j_0$. Since $A_j \cap (B' \times B'')$ is a pure $p$–dimensional analytic subset of $B' \times B''$, $\pi|_{A_j}$ is a branched covering onto $B'$ ([**35**], Chapter 1, Theorem 3.7). In particular $\pi|_{A_j}^{-1}(0) \neq \emptyset$ and hence $A_j \cap 0 \times \mathbb{C}^{n-p} \neq \emptyset$, for all $j \geq j_0$. □

LEMMA 4.26. *Let $\Omega \subset \mathbb{C}^n$ be an open set and $A$, $(A_j)_{j \in \mathbb{N}}$ be pure $p$–dimensional analytic subsets of $\Omega$ such that $\lim A_j = A$ in $\mathcal{C}(\Omega)$. Let $B \subset \Omega$ be a pure $q$–dimensional analytic subset. Assume that $A \cap B \neq \emptyset$ and $A$ and $B$ are in relative general position. Then $A_j \cap B \neq \emptyset$ for all large enough $j$.*

PROOF. Let $x \in A \cap B$; since $A \cap B$ is of pure dimension $p + q - n$, there is an affine plane $H$ of dimension $2n - (p+q)$ through $x$ such that $x$ is isolated in $A \cap B \cap H$ ([**35**], 3.5, Proposition 1). Thus $(x, x, x) \in \Omega^3$ is isolated in $(A \times B \times H) \cap \Delta_3$, where $\Delta_3 = \{(z, z, z) | z \in \mathbb{C}^n\}$. Moreover, $\lim(A_i \times B \times H) = A \times B \times H$ in $\mathcal{C}(\Omega^3)$. Lemma 4.25 implies that $(A_i \times B \times H) \cap \Delta_3 \neq \emptyset$ for $i$ large enough, in particular $A_i \cap B \neq \emptyset$. □

PROOF OF THEOREM 4.23. Passing to a subsequence we may assume that $(L_n)_{n \in \mathbb{N}}$ converges; let $L' = \lim L_n$. Since $L_n$ is closed for every $n$ and $\bigcup_n L_n$ is closed in $X \setminus L$, we must have $L' \subset L$. Furthermore $L'$ coincides with the set of points in $L$ which are essentially singular to $\bigcup_n L_n$. Therefore (cf. [**127**]) $L'$ is a union of irreducible components of $L$. If $L' \neq L$, there exists an irreducible component $L''$ of $L$ with $L'' \not\subset L'$ and $L' \cap L'' \neq \emptyset$. Since $L'$ and $L''$ are in relative general position, Lemma 4.26 implies $L_j \cap L'' \neq \emptyset$ for all $j$ big enough. This is a contradiction and thus $L' = L$. □

**4.4. The proof of Theorem 4.14.** Since $\mathrm{Geo}(X) < \infty$ and $\mathcal{H}^1_{(2),\mathrm{ex}}(X) = \mathcal{H}^1_{(2)}(X) \neq 0$ the assumptions of Scholium 4.18 are satisfied. Let $\mathcal{R}$ be the canonical equivalence relation on $X$, fix a closed holomorphic square–integrable 1–form $\omega$ defining $\mathcal{R}$ and choose $f : X \to \mathbb{C}$ holomorphic with $\omega = df$.

*Claim:* All leaves of $\mathcal{R}$, or equivalently all level sets of $f$, are compact.

The vanishing locus $A$ of $df$ is analytic, and $f$ is constant on its connected components; hence $f(A) \subset \mathbb{C}$ is countable. The set $f(X \setminus A)$ in open and non–empty. Therefore $f(X) \subset \mathbb{C}$ is measurable of positive measure, and Sard's theorem implies that for almost every $\rho \in f(X)$, $f^{-1}(\rho)$ is a submanifold of $X$. Therefore, the coarea integral $\int \mathrm{Vol}\, f^{-1}(\rho) d\mathcal{L}(\rho)$, where $d\mathcal{L}(\rho)$ is the volume form of the standard Kähler metric on $\mathbb{C}$, is well–defined and equals $\|df\|_2^2 < \infty$. Hence, for almost every $\rho \in f(X)$ the fiber $f^{-1}(\rho)$ is a finite volume submanifold of $X$ and hence compact since $\mathrm{Geo}(X) < \infty$ (Proposition 4.19).

The set

$$U' = \{x \in X | x \text{ belongs to a compact level set of } f\}$$

is therefore not empty. It is open, as follows from the following Lemma in general topology.

LEMMA 4.27 (see [128], Hilfssatz 3). *Let $f\colon A \to B$ be a continuous map, where $A$ and $B$ are locally compact spaces. Then*
$$U' = \{x \in A \,|\, x \text{ belongs to a compact level set of } f\}$$
*is open.*

Let $U$ be any connected component of $U'$. We now show that $U$ is closed. The function $f|_U \colon U \to \mathbb{C}$ is holomorphic with compact level sets. Applying Stein factorisation, we obtain a commutative diagram

$$\begin{array}{ccc} U & \xrightarrow{f} & \mathbb{C} \\ {\scriptstyle \pi}\downarrow & \nearrow & \\ S & & \end{array}$$

where $S$ is a Riemann surface, and the map $\pi$ is open, holomorphic, proper and its fibers are the level sets of $f|_U$. Observe that the image $\pi(C)$ of the set $C$ of critical points of $\pi$ is discrete. The Riemann surface $S_1 = S \setminus \pi(C)$ is connected and
$$\pi\colon U_1 = \pi^{-1}(S_1) \longrightarrow S_1$$
is a smoothly locally trivial fiber bundle. In particular, if $K$ denotes the Kähler form on $X$, it follows from Stokes's theorem that the function
$$S_1 \longrightarrow \mathbb{R}$$
$$s \longmapsto \int_{\pi^{-1}(s)} K^{n-1}$$
is constant, say equal to $v$. Hence $\mathrm{Vol}\,(\pi^{-1}(s)) = v$, for all $s \in S_1$, equivalently, $\mathrm{Vol}\,(L(z)) = v$, for all $z \in U_1$.

Let $x \in \bar{U}$; since $\bar{U} = \bar{U}_1$ we can find a sequence $(x_n)_{n\in\mathbb{N}}$ in $U_1$ such that $\lim x_n = x$. Since $L(x_n)$ is connected and $\mathrm{Vol}\,(L(x_n)) = v$, $\forall n \in \mathbb{N}$, there exists $R > 0$ (Proposition 4.19 2.) such that $\overline{\bigcup_n L(x_n)} \subset \overline{B}(x, R)$. This implies that any convergent subsequence of $(L(x_n))_{x\in\mathbb{N}}$ converges uniformly and hence has a connected limit.

Let $z \in \overline{\bigcup_n L(x_n)} \setminus \bigcup_n L(x_n)$ and $z_{n_k} \in L(x_{n_k})$, with $\lim z_{n_k} = z$. Passing to a subsequence we may assume that $(L(x_{n_k}))_{k\in\mathbb{N}}$ converges to a limit $E$, which is connected. Clearly, $\{x, z\} \subset E$ and $E \subset f^{-1}(f(x))$, and thus $E \subset L(x)$. This shows that $\overline{\bigcup_n L(x_n)} \subset U \cup L(x)$, and hence that $\bigcup_n L(x_n)$ is an analytic subset of $X \setminus L(x)$, since $\bigcup_n L(x_n)$ is analytic in $U$. It follows then from Lemma 4.24 and Theorem 4.23 that some subsequence of $(L(x_n))_{n\in\mathbb{N}}$ converges to $L(x)$, which implies that $L(x) \subset \overline{B}(x, R)$ and therefore $L(x)$ is compact. This shows the claim that $U$ is closed, hence $U = X$. The map $\pi\colon X \to S$ is now the global Stein factorisation of $f$; in particular its fibers are permuted by $\mathrm{Aut}\, X$ since the levels of $f$ are given by the canonical equivalence relation on $X$.

Since all fibers of $\pi$ have the same volume, we have $\pi^*(\mathcal{H}^1_{(2)}(S)) \subset \mathcal{H}^1_{(2)}(X)$, and $\pi^*\colon \mathcal{H}^1_{(2)}(S) \to \mathcal{H}^1_{(2)}(X) = \mathcal{H}^1_{(2),\mathrm{ex}}(X)$ is clearly injective. Let $\alpha \in \mathcal{H}^1_{(2)}(X) = $

$\mathcal{H}^1_{(2),\mathrm{ex}}(X)$ and $g\colon X \to \mathbb{C}$ be harmonic (see Lemma 4.12) with $dg = \alpha$. Since $X$ is Kähler, $g$ is pluriharmonic by Theorem 6.1, and its restriction to any fiber of $\pi$ is therefore constant. Thus $g$ factors through $\pi$, which shows that $\pi^*$ is surjective. This shows that $\mathcal{H}^1_{(2),\mathrm{ex}}(S) \neq 0$, and therefore $S$ has to be hyperbolic and non–compact. Its fundamental group is therefore free on $r \in \mathbb{N} \cup \{\infty\}$ generators; on the other hand $H^1(X,\mathbb{R}) = 0$, hence $H^1(S,\mathbb{R}) = 0$, which implies $r = 0$ and $S = \mathbb{D}^2$.

This completes the proof of Theorem 4.14.

## 5. Fibering Kähler manifolds over Riemann surfaces

In this Section we present the result of Arapura, Bressler and Ramachandran [5] that an extension of a group with infinitely many ends by a finitely generated group is not Kähler. This will be a corollary of the following alternative version of Theorem 4.14:

THEOREM 4.28 ([5]). *Let $X$ be a complete Kähler manifold with $\mathrm{Geo}(X) < \infty$. Assume $\dim \mathcal{H}^1_{(2),\mathrm{ex}}(X) \geq 2$. Then there exists a hyperbolic Riemann surface $S$ and a proper holomorphic map*

$$h\colon X \longrightarrow S$$

*with connected fibers. Moreover,*

1. *the fibers of $h$ are permuted by $\mathrm{Aut}\, X$, and*
2. *the map $h$ induces an isomorphism*

$$h^*\colon \mathcal{H}^1_{(2),\mathrm{ex}}(S) \longrightarrow \mathcal{H}^1_{(2),\mathrm{ex}}(X)\ .$$

COROLLARY 4.29. *Let $\Gamma$ be a Kähler group and*

$$1 \longrightarrow K \longrightarrow \Gamma \longrightarrow G \longrightarrow 1$$

*an exact sequence where $\bar{H}^1(G, l^2(G)) \neq 0$ and where $K$ has finite first Betti number. Then $G$ is commensurable with the fundamental group of a compact Riemann surface of genus $g \geq 2$.*

PROOF. Let $M$ be a compact Kähler manifold with $\Gamma = \pi_1(M)$, and let $X$ be the regular covering associated to $K \hookrightarrow \Gamma$. Since $\bar{H}^1(G, l^2(G)) \neq 0$, the group $G$ is infinite, and by Corollary 4.11 $\dim \mathcal{H}^1_{(2),\mathrm{ex}}(X) \geq 1$. Assume $\dim \mathcal{H}^1_{(2),\mathrm{ex}}(X) = 1$, and take $\alpha \in \mathcal{H}^1_{(2),\mathrm{ex}}(X)$, $\alpha \neq 0$. As $G$ acts unitarily on $\mathcal{H}^1_{(2),\mathrm{ex}}(X)$, the function $x \to \|\alpha(x)\|$ is $G$–invariant and in $L^2(X)$, which is not possible since $G$ is infinite and acts properly discontinuously on $X$. Therefore $\dim \mathcal{H}^1_{(2),\mathrm{ex}}(X) \geq 2$ and we obtain $h\colon X \to S$ as in Theorem 4.28.

Since $K_{\mathrm{ab}}$ is of finite rank, we have $\dim H^1(X,\mathbb{R}) < \infty$ and thus $\dim H^1(S,\mathbb{R}) < \infty$, which implies that $S$, being non–compact, is the quotient of $\mathbb{D}^2$ by a free group on $r < \infty$ generators and $G$ acts properly discontinuously on $S$ with compact quotient. If $r \geq 2$, the group $\mathrm{Iso}(S)$ of isometries of $S$ is finite, and since $G$ is infinite this implies $r \leq 1$. If $r = 1$, any discrete subgroup of $\mathrm{Iso}(S)$ is virtually cyclic and in particular does not act cocompactly on $S$. Therefore, we get $r = 0$ and $S = \mathbb{D}^2$. □

COROLLARY 4.30. *If $\Gamma$ is an extension of a group with infinitely many ends by a finitely generated group, then $\Gamma$ is not Kähler.*

Let $B_n$ be the *braid group* on $n$ strands, given by the presentation
$$B_n = \langle \sigma_1, \ldots, \sigma_{n-1} \mid \sigma_{i+1}\sigma_i\sigma_{i+1} = \sigma_i\sigma_{i+1}\sigma_i \text{ for all } i = 1, \ldots, n-2 \rangle \ .$$
This presentation shows that $b_1(B_n) = 1$, so $B_n$ is not Kähler.

Define the *pure braid group* $P_n$ as the kernel of the natural homomorphism from $B_n$ to the symmetric group $S_n$. We have seen in Example 1.19 that a Kähler group can be contained with finite index in a non–Kähler group. Thus, the fact that $B_n$ is not Kähler does not imply the same conclusion for $P_n$. One can calculate
$$b_1(P_n) = \frac{1}{2}n(n-1) \ ,$$
so $P_n$ is not Kähler for $n \equiv 2, 3 \pmod{4}$. More generally:

EXAMPLE 4.31 (Arapura [**4**]). The pure braid group $P_n$ is not Kähler for any $n$. To see this, one starts with the alternative presentation $B_3 = \langle a, b | a^3 = b^2 \rangle$. The subgroup $N = \langle a^3 \rangle$ is normal, and the quotient $B_3/N \simeq \mathbb{Z}_2 * \mathbb{Z}_3$ has infinitely many ends. Thus, we have an exact sequence
$$1 \longrightarrow N \cap P_3 \longrightarrow P_3 \longrightarrow G \longrightarrow 1 \ ,$$
where $N \cap P_3$ is finitely generated, and $G$ has infinitely many ends because it is contained with finite index in $\mathbb{Z}_2 * \mathbb{Z}_3$.

Now the natural homomorphism $p \colon P_n \to P_3$ has a finitely generated kernel, so we have
$$1 \longrightarrow p^{-1}(N \cap P_3) \longrightarrow P_n \longrightarrow G \longrightarrow 1 \ ,$$
where again $p^{-1}(N \cap P_3)$ is finitely generated. By Corollary 4.30, this shows that $P_n$ is not Kähler.

Turning now to the proof of Theorem 4.28, we recall the following general fact due to T. Napier, cf. [**97**].

LEMMA 4.32. *Let $X$ be a connected complex manifold and $\omega_1$ and $\omega_2$ linearly independent closed holomorphic 1–forms with $\omega_1 \wedge \omega_2 \equiv 0$. Then $h = \omega_2/\omega_1$ is a meromorphic function on $X$ without points of indeterminacy, i.e., a holomorphic mapping $X \to \mathbb{C}P^1$.*

PROOF. This is a version of the Castelnuovo–de Franchis Theorem 2.7, where however the manifold $X$ is not assumed compact, and $h$ is not a priori proper. Thus the argument given in Chapter 2 does not apply, and we have to proceed more directly.

Let $N \subset M \times \mathbb{C}P^1$ be the graph of the meromorphic function $h \colon M \to \mathbb{C}P^1$, and let $p \colon N \to M$ be the projection onto $M$. Let $g \colon N \to \mathbb{C}P^1$ be the projection onto $\mathbb{C}P^1$, and let $I$ be the set of points of indeterminacy. If $I$ is nonempty, then $I$ is an analytic set of pure dimension $n-2$ (see [**64**]). Moreover, $I \times \mathbb{C}P^1 \subset N$ and $p$ maps $N \setminus (I \times \mathbb{C}P^1)$ isomorphically onto $M \setminus I$. Let $J$ be a connected component of $I$ and, for each $\zeta \in \mathbb{C}P^1$, let $A_\zeta$ be the connected component of the fiber $g^{-1}(\zeta) = (M \times \{\zeta\}) \cap N$ (i.e. the level set of $g$) containing $J \times \{\zeta\}$. Since $p \colon N \to M$ is a proper map, the image $B_\zeta \equiv p(A_\zeta)$ is a connected analytic subset of $M$ containing $J$. Since $A_\zeta$ has pure dimension $n-1$ while $I \times \{\zeta\}$ has pure dimension $n-2$, and since the restriction of $p$ maps $N \setminus (I \times \mathbb{C}P^1)$ isomorphically onto $M \setminus I$, the set $B_\zeta$ has pure dimension $n-1$. Moreover, $B_\zeta \cap B_\xi \subset I$, for every pair of *distinct* points $\zeta, \xi \in \mathbb{C}P^1$, since $A_\zeta \cap A_\xi = \emptyset$.

Now fix a point $x_0 \in J \cap \bigcap_{\zeta \in \mathbb{C}P^1} B_\zeta$ and let $U$ be a neighbourhood of $x_0$ on which there exist holomorphic functions $f_1$ and $f_2$ such that $f_1(x_0) = f_2(x_0) = 0$ and $df_1 = \omega_1$ and $df_2 = \omega_2$. For each point $\zeta \in \mathbb{C}P^1$, let $C_\zeta$ be the irreducible component of $B_\zeta \cap U$ containing $x_0$. Then $f_2$ vanishes on $C_\zeta$ (as does $f_1$). For if $x_1$ is a nonsingular point of $C_\zeta$ in $U \setminus I$ and $(df_2)_{x_1} = (\omega_2)_{x_1} \neq 0$, then there exists a disk $D$ in $\mathbb{C}$ and a coordinate neighbourhood $(V, z)$ of $x_1$ in $U \setminus I$ such that $z = (z_1, \ldots, z_n)$ maps $V$ onto the polydisk $D^n$ in $\mathbb{C}^n$ and $z_1 = f_2|_V$. Since $0 \equiv df_1 \wedge df_2 = df_1 \wedge dz_1$, we have $f_1(z) = f_1(z_1)$ on $V$. Hence $h(z) = f_1'(z_1)$ on $V \subset U \setminus I$, and the connected component of $(h|_V)^{-1}(\zeta) = (h|_V)^{-1}(h(x_1))$ containing $x_1$ is the set $\{f_2(x_1)\} \times D^{n-1}$. In particular, since this set lies in $C_\zeta$ and $x_1$ is a nonsingular point of $C_\zeta$, we get

$$\omega_2(v) = df_2(v) = dz_1(v) = 0 \quad \forall v \in T_{x_1} C_\zeta \, .$$

This is also the case if $x_1$ is a nonsingular point of $C_\zeta$ at which $\omega_2 = 0$. Therefore, since $I \cap C_\zeta$ and $(C_\zeta)_{\text{sing}}$ are nowhere dense in $C_\zeta$, $f_2$ is constant on $C_\zeta$.

Thus $\{C_\zeta\}_{\zeta \in \mathbb{C}P^1}$ is a collection of *distinct* irreducible analytic sets of pure dimension $n-1$ in $U$ which lie in the proper analytic set $\{x \in U | f_2(x) = 0\}$ and which contain the point $x_0$. This clearly contradicts the local finiteness of the irreducible components of an analytic set. Hence $I = \emptyset$ and $h \colon M \to \mathbb{C}P^1$ is a holomorphic mapping. $\square$

PROOF OF THEOREM 4.28. Let $\mathcal{R}$ be the canonical equivalence relation on $X$ given by Scholium 4.18. Choose functions $f_1, f_2 \colon X \to \mathbb{R}$ such that $df_1, df_2$ are linearly independent, $df_i \in \mathcal{H}^1_{(2),\text{ex}}(X)$, $i = 1, 2$, and let

$$df_i = \omega_i + \bar{\omega}_i$$

be their decomposition into Hodge types. Let $h = \omega_2/\omega_1$ be the holomorphic function defined on $U = \{x \in X | \omega_1(x) \neq 0\}$ and $\tilde{h} \colon X \to \mathbb{C}P^1$ the associated holomorphic mapping given by Lemma 4.32.

We claim that the level sets of $\tilde{h}$ coincide with the leaves of $\mathcal{R}$: for $y \in \mathbb{C}P^1$ fix a biholomorphism $i_y \colon \mathbb{C}P^1 \setminus \{y\} \to \mathbb{C}$, and define $h_y = i_y \circ \tilde{h} \colon X \setminus \tilde{h}^{-1}(y) \to \mathbb{C}$. Since $\omega_2$ is closed and $h\omega_1 = \omega_2$ on $U$, we have $dh \wedge \omega_1 \equiv 0$, and hence $dh_y \wedge \omega_1 \equiv 0$ on $U \setminus \tilde{h}^{-1}(y)$ which implies that $dh_y \wedge \omega_1 \equiv 0$ on $X \setminus \tilde{h}^{-1}(y)$; hence the foliations induced on $X \setminus \tilde{h}^{-1}(y)$ by $\mathcal{R}$ and by the level sets of $\tilde{h}$ coincide for all $y \in \mathbb{C}P^1$. This shows the claim.

We show now that $\mathcal{R}$ has at least one compact leaf.

First we observe that every leaf of $\mathcal{R}$ is contained in a level set of the map

$$\Phi \colon X \longrightarrow \mathbb{R}^2$$
$$x \longmapsto (f_1(x), f_2(x)).$$

Indeed, on some connected neighbourhood $B$ of any point, we have $\omega_i = dg_{i,B}$ with $g_{i,B} \colon B \to \mathbb{C}$ holomorphic; in particular $f_i|_B$ and $2 \operatorname{Re} g_{i,B}$ coincide up to a constant, which implies that $\Phi|_B$ is constant on every leaf of the restricted foliation $\mathcal{R}|_B$, thus establishing the above observation. The restrictions to the open set

$$U_2 = \{x \in X | \operatorname{rank}_x \Phi = 2\}$$

of $\mathcal{R}$ and the foliation $\mathcal{R}_\Phi$ by levels sets of $\Phi$ give foliations of $U_2$ by real codimension 2 submanifolds, and the above observation implies then $\mathcal{R}|_{U_2} = R_\Phi|_{U_2}$.

Notice also that since $h\colon U \to \mathbb{C}$ is holomorphic non–constant,
$$U_2 = \{x \in X\mid d_x f_1 \wedge d_x f_2 \neq 0\}$$
is open and dense in $X$.

The set $\Phi(U_2) \subset \mathbb{R}^2$ is then open and non–empty, and Sard's theorem implies that for almost all $z \in \Phi(U_2)$, $\Phi^{-1}(z) \subset U_2$, and hence each of its connected components is a leaf of $\mathcal{R}$. Furthermore, the coarea formula and Cauchy–Schwarz imply
$$\int_{\mathbb{R}^2} \operatorname{Vol}(\Phi^{-1}(z)) d\mathcal{L}(z) \leq 2(\|df_1\|_2^2 + \|df_2\|_2^2) < \infty \ .$$
Hence, for almost every $z \in \Phi(U_2)$, $\Phi^{-1}(z)$ is of finite volume, and each of its connected components is a leaf of $\mathcal{R}$. Since $\operatorname{Geo}(X) < \infty$, Proposition 4.19 implies that $\mathcal{R}$ has at least one compact leaf. Thus the holomorphic map
$$\tilde{h}\colon X \longrightarrow \mathbb{C}P^1$$
has at least one compact level set. The argument then proceeds exactly as in the proof of Theorem 4.14. □

REMARK 4.33. From Sections 4 and 5 one can extract the following general facts: Let $X$ be a complex manifold and $U$ an isotropic subspace of *closed* holomorphic 1– forms for the cup product $H^0(\Omega_X^1) \times H^0(\Omega_X^1) \to H^0(\Omega_X^2)$, with $\dim U \geq 2$. Then the canonical equivalence relation $\mathcal{R}_U$ on $X$ associated to $U$ is given by the levels of a holomorphic map $h\colon X \to \mathbb{C}P^1$.

1. If $X$ is compact, Stein factorisation implies that $X$ fibers over a compact Riemann surface of genus $g \geq 2$ (since $\dim U \geq 2$). This is the Castelnuovo–de Franchis Theorem 2.7.
2. If $X$ is Kähler with $\operatorname{Geo}(X) < \infty$ and if $h$ has at least one compact level set, then all level sets are compact and $X$ fibers over a Riemann surface $S$.

CHAPTER 5

# Existence theorems for harmonic maps

This Chapter gives a brief discussion of some results needed in later chapters on existence and uniqueness of harmonic maps in homotopy classes of continuous maps with non–positively curved target manifolds. For more details we refer the reader to the original papers, in particular the one by Eells and Sampson [45], and to [44]. At the end of Chapter 6 a more general result will be needed, in a situation where the target space is not a manifold. This is discussed in [62] and in [82].

We are grateful to F. Labourie for his lecture on which this Chapter is based.

## 1. Definitions

Throughout this Chapter, we denote by $(M, g)$ a compact Riemannian manifold, and by $(N, h)$ a compact Riemannian manifold with non–positive sectional curvature.

DEFINITION 5.1. Let $f: M \to N$ be a smooth map.
(i) The *energy density* of $f$ is the function

$$e(f): M \longrightarrow \mathbb{R}^+$$
$$x \longmapsto \|df(x)\|_2^2 = \operatorname{Trace}(df_x^* df_x) = \sum_i \lambda_i \in \mathbb{R}^+,$$

where $\lambda_i$ are the eigenvalues of $(df_x^* df_x): T_x M \to T_x M$.
(ii) The *energy* of $f$ is the integral

$$E(f) = \int_M e(f) d\nu_M,$$

where $d\nu_M$ is the volume element of the metric $g$.

If $f_t: M \to N$ is a smooth 1–parameter family of maps, we have:

$$\frac{d}{dt} E(f_t)|_{t=0} = -\int_M \langle \Delta f, \frac{\partial f_t}{\partial t} \rangle d\nu_M.$$

Thus the Euler–Lagrange equation for the energy functional is $\Delta f = 0$, where $\Delta f = \operatorname{div} \nabla f = d^* df$ is the Laplace operator applied to $f$. This motivates the following definition.

DEFINITION 5.2. The map $f$ is *harmonic* if $\Delta f = 0$.

Thus, by definition, harmonic maps are critical points of the energy functional. They have the following property:

FACT 5.3. Let $f: M \to N$ be a harmonic map, and $\Psi: N \to \mathbb{R}$ a convex function. Then the function $\Psi \circ f: M \to \mathbb{R}$ is subharmonic.

## 2. Hartman's uniqueness theorem

The following is the basic uniqueness result for harmonic maps.

THEOREM 5.4 (Hartman (1967)). *Let $M$ be compact and $(N, h)$ have non–positive sectional curvature. If $f_0, f_1 \colon M \to N$ are two homotopic harmonic maps, then they are smoothly homotopic through harmonic maps.*

REMARK 5.5. In fact, one can obtain such a smooth homotopy for which all the paths $f_t(x)$ are parallel geodesics of length constant in $x$. Under these conditions, one says that the maps $f_0$ and $f_1$ are *parallel*.

PROOF. (Sketch) A *geodesic homotopy* is a 1–parameter family $f_t$ of maps such that the paths in $N$ given by $\gamma_v(t) = f_t(v)$ are geodesics for all $v \in M$, having velocities that may depend on $v$. If the target space has non–positive sectional curvature there are many geodesic homotopies; indeed, every homotopy is then homotopic relative to the ends to a geodesic homotopy. A functional on the space of maps from $M$ to $N$ is called *convex* if its restriction to every geodesic homotopy is a convex function.

The key ingredient of the proof is the following consequence of the second variation formula:

LEMMA 5.6. *The energy functional*
$$E \colon \mathrm{Map}(M, N) \longrightarrow \mathbb{R}$$
$$f \longmapsto E(f)$$
*is convex.*

Thus if $f_0$ and $f_1$ are homotopic harmonic maps, then both are minima of the energy functional. There is a geodesic homotopy connecting $f_0$ to $f_1$, and by the Lemma all maps $f_t$ in such a geodesic must be minima of the energy, and thus harmonic. □

COROLLARY 5.7. *If the curvature of $N$ is strictly negative, homotopic harmonic maps are equal as long as they do not factor through a closed geodesic.*

PROOF. In this case parallel segments $[f_0(x), f_1(x)]$ such as we may choose for the homotopy between two homotopic harmonic maps must all lie on the same geodesic. Therefore, $f_0 \sim f_1$ implies $f_0 = f_1$, unless the images are closed geodesics or points. □

## 3. The Eells–Sampson theorem

The following theorem is the most basic existence theorem for harmonic maps. All the other existence theorems mentioned in the book contain this one as a special case.

THEOREM 5.8 (Eells–Sampson [45]). *Let $M$ and $N$ be compact Riemannian manifolds, and assume that $N$ has non–negative sectional curvature. Then every homotopy class of continuous maps $M \to N$ contains a harmonic representative.*

This is proved by studying the solutions of the heat flow equation on $M \times [0, \infty)$. We pick an arbitrary smooth map $f_0 \colon M \to N$ in the given homotopy class, and use it as initial value for the flow

$$\tag{7} \frac{\partial f_t}{\partial t} = \Delta f_t \,.$$

This equation has the following properties:

THEOREM 5.9 (Eells–Sampson). *If $M$ and $N$ are compact and $N$ has non-positive sectional curvature, the following hold:*

(i) *The equation (7) has solutions for all $t > 0$.*

(ii) *The energy densities of the solutions are bounded, i.e., there exists a constant $K_0$ such that $e(f_t) < K_0$ for all $t > 0$ and all $x \in M$.*

(iii) *The norms of the derivatives $\frac{\partial f_t}{\partial t}$ are bounded, for example by a bound on the norms of derivatives of the initial map $f_0$.*

(iv) *The energy $E(t) = E(f_t)$ is decreasing.*

(v) $\lim_{t \to \infty} \|\Delta f_t\|^2 = 0$.

Given this result, it is easy to deduce Theorem 5.8 by extracting a convergent subsequence from the solutions of the heat flow (7). If one assumes that the sectional curvature of $(N, h)$ is strictly negative, then it is not necessary to pass to a subsequence, cf. [60].

PROOF. (of Theorem 5.9) We only sketch the proof; for a more complete account see [60] or [77].

(iv) and (v) are obtained straightforwardly by differentiating under the defining integral of $E$.

Next, to obtain (ii), one calculates the Laplacian of the energy density function $e$ on $M$ as follows:

$$\Delta e(f_t) = |\nabla df|^2 + \langle \text{Ric}^M \nabla_{v_i} f_t, \nabla_{v_i} f_t \rangle \\ - \langle K_N(df \cdot v_i, df \cdot v_j) df \cdot v_i, df \cdot v_j \rangle ,$$

where $v_1, \ldots, v_n$ form an orthonormal frame at $T_x M$. With the assumption $K_N \leq 0$, taking $R$ such that $\text{Ric}^M \geq -R$, we obtain the following Bochner inequality:

$$\frac{\partial e}{\partial t} \leq \Delta e + R e .$$

Applying a standard lemma on the growth of solutions of parabolic equations ([77], Lemma 2.3.1) we obtain a bound

$$e(f_t)(x) \leq c R^{-2} \sup_{y \in M} e(f_0)(y) ,$$

for some constant $c$.

Using Schauder estimates for parabolic equations, one checks that (ii) implies (iii). Finally, to show (i), one uses the fact that the heat flow equation is parabolic, so there exists a solution for small values of $t$. The bound of (iii) allows the extension of such solutions to arbitrary $t$. □

## 1. Equivariant harmonic maps

In later chapters we will occasionally use existence theorems for equivariant harmonic maps. We consider a compact, or at least complete, Riemannian manifold $M$, and a simply connected complete Riemannian manifold $V$, e.g. $V = \tilde{N}$, the universal cover of the $N$ considered before. Given a representation

$$\rho \colon \pi_1(M) \longrightarrow \text{Iso}(V) ,$$

we are looking for a $\rho$–twisted, or $\rho$–equivariant, harmonic map

$$f \colon \tilde{M} \longrightarrow V .$$

The $\rho$–equivariance means that $f(\gamma v) = \rho(\gamma)(f(v))$ for all $\gamma \in \pi_1(M), v \in V$. This is the same as a harmonic section of the flat bundle over $M$ with fiber $V$ associated with $\rho$, where harmonicity of a section is defined by extremising a suitable notion of vertical energy. If the sectional curvature of $V$ is non–positive, $V$ is contractible, so that the bundle has a continuous section. Applying the heat flow essentially as in the proof of the Eells–Sampson Theorem 5.8, one can give a criterion for the existence of an equivariant harmonic map. To state this criterion, we use the following definition:

DEFINITION 5.10. Let $V$ be a Riemannian manifold. A subgroup $\Gamma \subset \operatorname{Iso}(V)$ is called *reductive* if there exists a convex set $C \subset V$ preserved by $\Gamma$ and such that:

(i) The set $C$ is the Riemannian product $C = C_1 \times E$, where $C_1$ is a convex set, and $E$ is a Euclidean space.

(ii) The subgroup $\Gamma$ preserves the decomposition, and its restriction $\rho_1 \colon \Gamma \to \operatorname{Iso}(C_1)$ does not fix any point at infinity in $C_1$.

This definition extends the classical concept of reductive groups in the following sense:

LEMMA 5.11. *Suppose $G$ is real algebraic group, and $V = G/K$ a homogeneous manifold. Then a subgroup $\Gamma \subset \operatorname{Iso}(V) = G$ is reductive (in the sense of Definition 5.10) if and only if its Zariski closure is a reductive subgroup of $G$ (in the usual sense).*

The extension of the Eells–Sampson theorem is:

THEOREM 5.12. *Let $M$ be a compact Riemannian manifold, and $V$ a simply connected complete Riemannian manifold with non–positive sectional curvature. Let $\rho \colon \pi_1(M) \to \operatorname{Iso}(V)$ be a representation with reductive image $\rho(\pi_1(M))$. Then there exists a $\rho$–equivariant harmonic map*

$$f \colon \tilde{M} \to V.$$

This was proved by Corlette [36] for the case where $V$ is symmetric, using an argument related to the proofs of the existence of Hermitian–Einstein connections on stable bundles due to Donaldson and Uhlenbeck–Yau (compare Chapter 7). Donaldson [43] proved the case where $V$ is hyperbolic three–space, by adapting the heat flow method of Eells–Sampson. Labourie [87] clarified the definition of reductivity of the representation and pointed out that Donaldson's adaptation of the Eells–Sampson argument works in the general case.

REMARK 5.13. When the image $\rho(\pi_1(M))$ is contained in a cocompact isometry subgroup, Theorem 5.12 implies the Eells–Sampson Theorem 5.8.

PROOF. (of Theorem 5.12) The proof is an immediate consequence of the following two lemmata.

LEMMA 5.14. *Assume that $\rho(\pi_1(M))$ does not fix a point at infinity in $V$. Then there exist twisted harmonic maps.*

PROOF. As $V$ is simply connected, there exists a continuous twisted map $f_0 \colon \tilde{M} \to V$. Now consider the heat flow equation

$$\frac{\partial f_t}{\partial t} = \Delta f_t$$

with initial data $f_0$. Its solution also satisfies Theorem 5.9.

We cannot apply directly the Arzela–Ascoli argument as before to get a convergent subsequence, because given $t_n \to \infty$, $f_{t_n}(v)$ may go to infinity in $V$. Nevertheless, it can be shown that $f_t$ is uniformly Lipschitz, so there is a uniform bound
$$d(f_t(\gamma v), f_t(v)) < K_0 \ .$$
Therefore $d(\rho(\gamma) f_{t_n}(v), f_{t_n}(v)) < K_0$, and thus $f_{t_n}(v)$ cannot go to infinity for any $v \in V$, otherwise $\rho(\pi_1(M))$ would fix the corresponding infinity point.

Consequently we can apply Arzela–Ascoli and obtain a harmonic map. □

LEMMA 5.15. *Assume that $V$ is Euclidean, then there exist harmonic maps.*

The proof of this lemma parallels that of the previous one. □

There is a converse to Theorem 5.12:

THEOREM 5.16 (Corlette [36], Labourie [87]). *Assume that there are no flat half strips in $V$. If there exists a $\rho$-equivariant harmonic map, then $\rho(\pi_1(M))$ is reductive.*

The technical hypothesis excluding flat half–strips is always satisfied if $V$ is real analytic, or if its sectional curvature is strictly negative.

CHAPTER 6

# Applications of harmonic maps

## 1. Existence of pluriharmonic maps

In the previous Chapter we have studied the existence of harmonic maps between Riemannian manifolds, under the assumption of non–positive sectional curvature in the target space (Theorem 5.8).

The theory of harmonic maps with domain a compact Kähler manifold is much richer than the theory of harmonic maps with domain a general Riemannian manifold. This situation is analogous to that in the theory of harmonic forms, which is richer and has more profound applications on Kähler manifolds than on general Riemannian manifolds. For harmonic maps this richer structure is made precise by the following theorem. This Section shall be devoted to its proof, and the following Sections to various applications.

THEOREM 6.1 (Siu [**120**], Sampson [**109**]). *Let $M$ be a compact Kähler manifold, and $N$ a Riemannian manifold with non–positive Hermitian sectional curvature. Then every harmonic map $f \colon M \to N$ is pluriharmonic.*

We first explain the terminology used in the Theorem. We use the universal symbol $R$ to denote the Riemannian curvature $(0,4)$–tensor of $N$, or the corresponding $(1,3)$–tensor, related by the equation $R(X,Y,Z,T) = g(R(X,Y)Z,T)$. Notational conventions are chosen so that if $X, Y$ is an orthonormal basis for a plane in $T_xN$, then $R(X,Y,X,Y)$ is the sectional curvature of that plane. The tensor $R$ may be extended to the complexified tangent bundle $TN \otimes \mathbb{C}$. The *Hermitian sectional curvature* is the Hermitian form on $\Lambda^2 TN \otimes \mathbb{C}$ sending $X \wedge Y$ to $R(X,Y,\bar{X},\bar{Y}) \in \mathbb{R}$. If the vector fields $X, Y$ are real and orthonormal, this is just the sectional curvature of the plane they determine, but in general the condition that $R(X,Y,\bar{X},\bar{Y}) \leq 0$ is stronger than the condition of non–positive sectional curvature.

Let us recall that a map $f \colon M \to N$, with $M$ a complex manifold, is *pluriharmonic* if for every germ of complex curve $C \subset M$ the restriction $f \colon C \to N$ is harmonic. Note that, since harmonicity of maps of Riemann surfaces is independent of the Riemannian metric, and depends only on the complex structure of the surface (compatible with the metric), the condition that $f$ be pluriharmonic depends on the complex structure of $M$ but not on its Riemannian metric. Although the two concepts are not comparable in general, pluriharmonic maps are harmonic when $M$ is a Kähler manifold.

The usefulness of Theorem 6.1, as clearly explained in Siu's original paper [**120**], is that to produce harmonic maps, one must solve a determined system of equations, so there is a reasonable chance of obtaining solutions (as in the Eells–Sampson Theorem 5.8). But the system of equations that produces pluriharmonic maps is overdetermined, and thus is unlikely to have solutions except under exceptional circumstances. Thus the existence of pluriharmonic maps imposes strong restrictions on the source space $M$ and the possible homotopy classes of maps with domain $M$.

A twisted version of Theorem 6.1 holds for $\rho$–equivariant harmonic maps $f\colon \tilde{M} \to N$, where $\tilde{M}$ is the universal covering of $M$ and $\rho\colon \pi_1(M) \to Iso(N)$ is a reductive representation, namely equivariant harmonic maps are pluriharmonic. Since the proof of the twisted version is identical to the proof of Theorem 6.1, for simplicity we restrict ourselves to the proof of the untwisted case.

As applications of these theorems, restrictions are established on the homotopy classes of maps $f\colon M \to N$, and on the existence of representations $\rho\colon \pi_1(M) \to Iso(N)$. In particular, in the case of a symmetric space $N = G/K$, restrictions on the representations $\pi_1(M) \to G$ are found.

TERMINOLOGY 6.2. We will use the following notations:
(i) $\nabla$ is the Levi-Civita connection on $TN$, and on $f^*TN$.
(ii) $d_\nabla \colon A^k(M, f^*TN) \to A^{k+1}(M, f^*TN)$ is the covariant derivative given by $d_\nabla(\alpha \otimes s) = d\alpha \otimes s + (-1)^k \alpha \otimes \nabla s$.
(iii) $R$, defined by $Rs = -d_\nabla^2 s$, is the curvature.

**Warning:** In this Section the symbol "=" will sometimes mean equality up to a universal non–zero multiplicative constant. If an expression on either side of the "=" sign consists of more than one term, there could be a *different* but universal non–zero multiplicative constant in front of *each* term.

EXAMPLE 6.3. For any $\alpha \in T^*M$, here is an equality (*up to a multiplicative constant!*)
$$*\alpha = \omega^{n-1} \wedge J\alpha .$$

We can use this identity to replace the Hodge $*$–operator in the definition of the Laplacian, and the condition for harmonicity for $f\colon M \to N$, $\Delta f = 0$, can be rewritten as
$$\Delta f = *d_\nabla(\omega^{n-1} \wedge J(df)) = *(\omega^{n-1} \wedge d_\nabla(J(df))) = 0$$

As the Hodge operator is bijective, we reach the following characterisation:

FACT 6.4. Let $M$ be a compact Kähler manifold. A map $f\colon M \to N$ is harmonic if and only if
$$\omega^{n-1} \wedge d_\nabla d^c f = 0$$

Pluriharmonicity admits a similar characterisation:

FACT 6.5. Let $M$ be compact Kähler. A map $f\colon M \to N$ is pluriharmonic if and only if
$$d_\nabla d^c f = 0$$

This characterisation of pluriharmonicity follows immediately from the fact that, on a curve, i.e., when $n = 1$, the harmonic equation is equivalent to $d_\nabla d^c f = 0$. Thus the equation $d_\nabla d^c f = 0$ restricts, on each complex curve, to the harmonic equation on that curve.

The previous characterisations show that pluriharmonic maps from compact Kähler manifolds are harmonic. We proceed now to prove the Siu–Sampson Theorem 6.1, which is the converse result when $N$ is Hermitian non–positively curved.

*Case 1: Flat target space.* Let $f\colon M \to S^1$ be a smooth map (or analogously $f\colon \tilde{M} \to \mathbb{R}$ an equivariant map with respect to a representation of $\pi_1(M)$ in the group of translations of $\mathbb{R}$). Then $df$ is an ordinary 1–form on $M$ (or descends to an ordinary 1–form on $M$ in the equivariant case) and $f$ is harmonic if and only if $df$ is a harmonic 1–form.

The form $dd^c f \wedge dd^c f \wedge \omega^{n-2} = d(d^c f \wedge dd^c f \wedge \omega^{n-2})$ is exact, so by Stokes's theorem we have that
$$\int_M dd^c f \wedge dd^c f \wedge \omega^{n-2} = 0$$
The next step will be to show this form is non–positive:

*Claim:* If $f$ is harmonic, then $dd^c f \wedge dd^c f \wedge \omega^{n-2} \leq 0$, and it equals zero if and only if $dd^c f = 0$.

This claim is a consequence of the statement at the end of the following paragraph, which is often called the *Hodge signature theorem* since a classical application is the computation of the signature of an algebraic surface in terms if its Hodge decomposition.

Let $V$ be a real vector space with compatible almost complex structure $J$ and inner product $\langle .,.\rangle$. (This will be $T_p^*M$ in our case.) There is a natural orthogonal decomposition as $U(V)$–modules
$$\Lambda^{1,1}V = \mathbb{R}\omega \oplus (\Lambda^{1,1}V)_0 \,,$$
where $\omega$ is the fundamental form. This decomposition corresponds to the decomposition of the space of Hermitian endomorphisms of $V$ into the direct sum of multiples of the identity and traceless endomorphisms. The component of traceless endomorphisms, $(\Lambda^{1,1}V)_0$, is an irreducible $U(V)$–module, and one can check that the $U(V)$–invariant pairing

(8) $$\alpha \otimes \beta \longmapsto \alpha \wedge \beta \wedge \omega^{n-2}$$

is not identically zero. Therefore, it is definite, and computation of any example shows that it is negative definite.

Since $f$ is harmonic, by Fact 6.4 $\omega^{n-1} \wedge dd^c f = 0$, which means that $dd^c f$ belongs to $(\Lambda^{1,1}V)_0$. The pairing (8) on $(\Lambda^{1,1}V)_0$ is negative definite, thus the form $dd^c f \wedge dd^c f \wedge \omega^{n-2}$ is pointwise non–positive on $M$ and has integral 0, so it must be zero everywhere. Hence the form $dd^c f$ must be zero, and by Fact 6.5 $f$ is pluriharmonic.

*Case 2: Arbitrary Hermitian nonpositively curved target space $N$.* In the general case we must repeat the above reasoning taking into account the curvature of the target space. Let us first define the following product of forms, composing the ordinary wedge product with the pairing of the coefficients given by the metric of $N$:

$$\langle .\wedge .\rangle : \Lambda^k(M, f^*TN) \otimes \Lambda^m(M, f^*TN) \xrightarrow{\wedge} A^{k+m}(M, f^*TN \otimes f^*TN)$$
$$\xrightarrow{\langle ...\rangle_N} A^{k+m}(M, \mathbb{R}) = A^{k+m}(M) \,.$$

The form $\langle d_\nabla d^c f \wedge d^c f\rangle$ is in $A^3(M)$. Since $\nabla$ is the Levi–Civita connection on $N$, we have:
$$d(\langle d_\nabla d^c f \wedge d^c f\rangle \wedge \omega^{n-2}) = \langle d_\nabla^2 d^c f \wedge d^c f\rangle \wedge \omega^{n-2} + \langle d_\nabla d^c f \wedge d_\nabla d^c f\rangle \wedge \omega^{n-2}$$
$$= \langle -R d^c f \wedge d^c f\rangle \wedge \omega^{n-2} + \langle d_\nabla d^c f \wedge d_\nabla d^c f\rangle \wedge \omega^{n-2} \,.$$

Now comes the only *global* step of the argument. Since $M$ is compact, this exact form integrates to zero, so

$$\int_M (\langle -Rd^cf \wedge d^cf \rangle \wedge \omega^{n-2} + \langle d_\nabla d^cf \wedge d_\nabla d^cf \rangle \wedge \omega^{n-2}) = 0 \ . \tag{9}$$

One can show that the second summand $\langle d_\nabla d^cf \wedge d_\nabla d^cf \rangle \wedge \omega^{n-2}$ is negative by repeating verbatim the above reasoning in the flat case, so what is left to prove, is:

$$\langle -Rd^cf \wedge d^cf \rangle \wedge \omega^{n-2} \leq 0 \ .$$

This is done by relating this form to Hermitian sectional curvature.

The form $\langle Rd^cf \wedge d^cf \rangle$ is in $A^4(M)$. It is obtained by skew–symmetrising the Riemannian curvature tensor $f^*R(.,.,J.,J.)$ defined by

$$(X_1, X_2, X_3, X_4) \longmapsto R(df(X_1), df(X_2), df(JX_3), df(JX_4)) \ .$$

This tensor is already skew–symmetric with respect to the first and second terms, and also with respect to the third and fourth terms, so we will denote it as $f^*R(X_1 \wedge X_2, JX_3 \wedge JX_4)$. Recall that $f^*R$ is then a *symmetric* bilinear form on $\Lambda^2(TM)$, and that it satisfies the Bianchi identity

$$f^*R(X_1 \wedge X_2, X_3 \wedge X_4) + f^*R(X_3 \wedge X_1, X_2 \wedge X_4) + f^*R(X_2 \wedge X_3, X_1 \wedge X_4) = 0 \ .$$

What we actually need is the wedge product of the skew–symmetrisation of this tensor with $\omega^{n-2}$. Now, if $\alpha$ is any four–form on $M$, then $\alpha \wedge \omega^{n-2} = a\omega^n$, and $a$ is computed as follows. Let $P$ be a subspace of $TM$ of real dimension four which is invariant under $J$, i.e., a complex two–dimensional subspace of $TM$. If $X, Y \in TM$ are vectors so that $X, Y, JX, JY$ form an orthonormal basis of $P$, then the number $\alpha(X, JX, Y, JY)$ depends only on $P$, and the coefficent $a$ is the average of these numbers over all four–dimensional $J$–invariant subspaces $P$. Thus we only need to show that

$$\langle -Rd^cf \wedge d^cf \rangle(X, JX, Y, JY) \leq 0$$

holds for all $X, Y \in TM$, equivalently, that

$$\langle Rd^cf \wedge d^cf \rangle(X, Y, JX, JY) \leq 0$$

holds for all $X, Y \in TM$. But this will follow from the non–positive Hermitian curvature assumption and the identity

$$\langle Rd^cf \wedge d^cf \rangle(X, Y, JX, JY) = R(Z, W, \bar{Z}, \bar{W}),$$

where $Z, W \in f^*TN \otimes \mathbb{C}$ are defined by $Z = df(X - iJX)$ and $W = df(Y - iJY)$.

This last identity can be checked by the following straightforward computation. First compute the above skew–symmetrisation by choosing any set of six representatives of the cosets of the symmetric group by the group of order four of skew–symmetries already present in $f^*R$, for example

$$\begin{aligned}\langle Rd^cf \wedge d^cf \rangle(X, Y, JX, JY) = {} & f^*R(X \wedge Y, J(JX \wedge JY)) \\ & - f^*R(X \wedge JX, J(Y \wedge JY)) \\ & + f^*R(X \wedge JY, J(Y \wedge JX)) \\ & + f^*R(Y \wedge JX, J(X \wedge JY)) \\ & - f^*R(Y \wedge JY, J(X \wedge JX)) \\ & + f^*R(JX \wedge JY, J(X \wedge Y)) \ .\end{aligned}$$

Denote these six terms, in the order listed, by $t_1, \cdots, t_6$.

Next, compute $R(Z, W, \bar{Z}, \bar{W})$, which is explicitly the following sum of eight terms:

$$\begin{aligned}
f^*R(X - iJX, Y - iJY, X + iJX, Y + iJY) &= f^*R(X \wedge Y, X \wedge Y) \\
&\quad - f^*R(JX \wedge JY, X \wedge Y) \\
&\quad - f^*R(X \wedge Y, JX \wedge JY) \\
&\quad + f^*R(X \wedge JY, JX \wedge Y) \\
&\quad + f^*R(JX \wedge Y, JX \wedge Y) \\
&\quad + f^*R(JX \wedge Y, X \wedge JY) \\
&\quad + f^*R(X \wedge JY, X \wedge JY) \\
&\quad + f^*R(JX \wedge JY, JX \wedge JY) \, .
\end{aligned}$$

Denote these eight terms, in the order listed, by $s_1, \cdots, s_8$.

Rewriting $t_1, \cdots, t_6$ by computing the second wedge–product argument using the identities $J(A \wedge B) = JA \wedge JB$ and $J^2 = -1$, we see that $s_1 = t_1$, $s_5 = t_4$, $s_7 = t_3$, and $s_8 = t_6$. Using the Bianchi identity, we see that $s_2 + s_4 = t_2$ and $s_3 + s_6 = t_5$. Therefore $\langle Rd^c f \wedge d^c f \rangle(X, Y, JX, JY) - R(Z, W, \bar{Z}, \bar{W}) \leq 0$, as claimed.

We have shown now that both summands in the integral (9) are non–positive. As the integral is zero, both must be identically zero. In particular, $\langle d_\nabla d^c f \wedge d_\nabla d^c f \rangle \wedge \omega^{n-2} = 0$. Since $f$ is harmonic, $d_\nabla d^c f$ lies in the bundle of traceless endomorphisms $(\Lambda T^*M)_0$. The pairing (8) is definite, so $d_\nabla d^c f = 0$, and $f$ is pluriharmonic as we wanted to show.

So far, we have clearly proved that there is a universal sign such that if the Hermitian sectional curvature of the target space has that sign, then harmonic maps are pluriharmonic. We have actually proved that Hermitian non-positive curvature implies that harmonic maps are pluriharmonic. But rather than attempting to convince the reader that the signs in the above argument are correct, we find it more convincing and conceptual to check the sign in the following example. We have taken care to discuss the example in a manner totally independent of the above calculations.

EXAMPLE 6.6. The natural projection

$$\pi \colon \mathbb{C}P^3 = (\mathbb{C}^4 \setminus \{0\})/\mathbb{C}^* \longrightarrow (\mathbb{H}^2 \setminus \{0\})/\mathbb{H}^* \cong \mathbb{H}P^1 \cong S^4,$$

defined by assigning to a complex line in $\mathbb{C}^4 \cong \mathbb{H}^2$ the unique quaternionic line containing it, is harmonic. The reason is that it is a Riemannian submersion whose fibers, namely the set of complex lines contained in a fixed quaternionic line, are complex projective lines, hence area–minimising 2–spheres in $\mathbb{C}P^3$. One knows from the beginning of the theory of harmonic maps (the original Eells-Sampson paper [45]) that Riemannian submersions with minimal fibers are harmonic.

The quotient $S^4$ has constant curvature $1 > 0$. If Theorem 6.1 held for non-negative sectional curvature, $\pi$ would be pluriharmonic. But $\pi$ is not pluriharmonic; Let $F$ be a fiber of $\pi$, (which is a complex projective line in $\mathbb{C}P^3$) and let $C$ be a complex projective line in $\mathbb{C}P^3$ which meets $F$ at exactly one point and is contained in an $\varepsilon$-neighbourhood of $F$. If $\pi$ were pluriharmonic, then $\pi|C$ would be a non–constant harmonic map of $C$ with image in an $\varepsilon$-neighbourhood of a point in $S^4$. But for small enough $\varepsilon$ this violates the maximum principle for harmonic maps [108].

Since there are harmonic maps into positively curved targets which are not pluriharmonic, our perhaps imprecise proof of Theorem 6.1, which only assured us that harmonic maps are pluriharmonic if the Hermitian sectional curvature has a certain fixed sign, now becomes precise and assures us that this sign must be non-positive, as stated.

## 2. First applications

We shall now study some of the consequences of the pluriharmonicity of harmonic maps (the Siu–Sampson theorem 6.1) in the case when the target space $N$ is a quotient by a lattice of a symmetric space $G/K$ of non–compact type. Here $G$ is a linear semi–simple Lie group without compact factors and $K$ is a maximal compact subgroup. Examples of such spaces $G/K$ are

$$SL(n,\mathbb{R})/SO(n), \quad SU(p,q)/S(U(p) \times U(q)), \quad SO_0(p,q)/(SO(p) \times O(q)) .$$

We recall briefly some of the basic algebraic and geometric facts we need concerning the spaces $G/K$. The Lie algebra $\mathfrak{g}$ of $G$ admits a Cartan decomposition

$$\mathfrak{g} = \mathfrak{k} \oplus \mathfrak{p}$$

satisfying

$$[\mathfrak{k},\mathfrak{k}] \subset \mathfrak{k}, \quad [\mathfrak{k},\mathfrak{p}] \subset \mathfrak{p}, \quad [\mathfrak{p},\mathfrak{p}] \subset \mathfrak{k}$$

and such that there is a natural isomorphism $T_{eK}(G/K) \cong \mathfrak{p}$.

The Killing form on $\mathfrak{g}$, which we will denote as $\langle .,. \rangle$, is a bilinear form which is positive definite on $\mathfrak{p}$, and negative definite on $\mathfrak{k}$. Therefore it defines an invariant metric on $G/K$, whose curvature tensor at $eK$ is

$$R(X,Y) = [X,Y].$$

The Hermitian sectional curvature on $T(G/K)^{\mathbb{C}} \cong \mathfrak{p}^{\mathbb{C}}$ is

$$R(X,Y,\bar{X},\bar{Y}) = \langle [X,Y],[\bar{X},\bar{Y}] \rangle \leq 0,$$

the last inequality is due to the fact that $[X,Y]$ belongs to the complexified subalgebra $\mathfrak{k}^{\mathbb{C}}$, where the Killing form is negative definite. Therefore the Riemannian manifolds $N = \Gamma\backslash G/K$ have non–positive Hermitian sectional curvature, and the Siu–Sampson Theorem 6.1 applies.

Let us recall that in the proof of that theorem we showed that if $f\colon M \to N$ is a harmonic map of a compact Kähler manifold to a manifold of non–positive Hermitian sectional curvature, then

  (i) $d_\nabla d^c f = 0$, i.e. $f$ is pluriharmonic.
  (ii) $\langle R(df(X), df(Y))df(\bar{X}), df(\bar{Y})\rangle = 0$ for all $X,Y \in T^{1,0}M$.

We will now give an interpretation of (ii) when $\tilde{N}$ is a symmetric space $G/K$. First, since $R(X,Y) = [X,Y]$, we see that

$$\langle R(X,Y)\bar{X},\bar{Y}\rangle = \langle [X,Y],\overline{[X,Y]}\rangle.$$

It follows that (ii) is equivalent to

  (iii) $R(df(X), df(Y)) = 0$ for all $X,Y \in T^{1,0}M$.

## 2. FIRST APPLICATIONS

This in turn has the following interpretation: Let $d''_\nabla$, also frequently written as $\bar{\partial}_\nabla$, denote the composition

$$d''_\nabla : A^{0,k}(M, f^*TN^{\mathbb{C}}) \hookrightarrow A^k(M, f^*TN^{\mathbb{C}})$$
$$\xrightarrow{d_\nabla} A^{k+1}(M, f^*TN^{\mathbb{C}}) \xrightarrow{\pi} A^{0,k+1}(M, f^*TN^{\mathbb{C}}) \ .$$

Property (iii) above is equivalent (by complex conjugation) to $R(df(X), df(Y)) = 0$ for all $X, Y \in T^{0,1}M$, which is equivalent to

(10) $$(d''_\nabla)^2 = 0 \ ,$$

the integrability condition that allows us to define a complex structure, called the *Koszul–Malgrange complex structure*, on $f^*TN^{\mathbb{C}}$ as follows: a local section $s$ of $f^*TN^{\mathbb{C}}$ is holomorphic if and only if $d''_\nabla s = 0$. We refer to [**83**] for the original proof that this integrability condition insures local trivialisations of $f^*TN^{\mathbb{C}}$ by local holomorphic sections.

Denote by $d'f$ the restriction of $df$ to $T^{1,0}M$. Property (i) above implies that

$$d''_\nabla d'f = 0 \ ,$$

and this means that $d'f$ is a holomorphic section of $T^{1,0*}M \otimes f^*TN^{\mathbb{C}}$, with the complex structure induced by that of $TM$ and the Koszul–Malgrange complex structure.

With these interpretations of pluriharmonicity understood, we are now ready to introduce, following Sampson [**109**], the main algebraic ingredient in studying pluriharmonic maps to locally symmetric targets. Let $f: M \to N$ be a harmonic (hence pluriharmonic) map from a compact Kähler manifold to a symmetric space of non–compact type. Set $\mathfrak{a}_x = df(T^{1,0}_x M) \subset T_{f(x)}N^{\mathbb{C}}$. By the natural identification of $T_{f(x)}N$ with $\mathfrak{p}$, we have an inclusion $\mathfrak{a}_x \subset \mathfrak{p}^{\mathbb{C}}$. Theorem 6.1 holds, so by Property (iii) the space $\mathfrak{a}$ is $R$–isotropic. As the curvature $R$ is the ordinary Lie bracket, it turns out that $\mathfrak{a}_x = df(T^{1,0}_x M)$ must be an Abelian subalgebra of $\mathfrak{g}$ contained in $\mathfrak{p}$, which we will denote simply by $\mathfrak{a}$. We explain its significance by first looking at two extreme cases.

### 2.1. Real hyperbolic space.
This is the extreme case where all Abelian subalgebras are small.

Set $G/K = H^n_{\mathbb{R}}$, the real hyperbolic space. It has constant scalar curvature $= -1$, and sectional curvature given by $R(X,Y) = X \wedge Y \in \Lambda^2 T_x N \cong SkewEnd(T_x N)$, where $X \wedge Y$ corresponds to the skew–symmetric endomorphism

$$X \wedge Y : T_x N \longrightarrow T_x N$$
$$Z \longmapsto \langle X, Z \rangle Y - \langle Y, Z \rangle X \ .$$

Now $\mathfrak{a} \subset \mathfrak{p}^{\mathbb{C}}$ being isotropic means that $X \wedge Y = 0$ for all $X, Y \in \mathfrak{a}$. This happens if and only if $\dim \mathfrak{a} \leq 1$. The complexified derivative

$$T_x M^{\mathbb{C}} = T^{1,0}_x M \oplus T^{0,1}_x M \xrightarrow{df} T_x N^{\mathbb{C}}$$

has image $\mathfrak{a} + \bar{\mathfrak{a}}$. Looking at the real tangent bundle we find an inclusion $df(T_x M) \otimes \mathbb{C} \subseteq \mathfrak{a} + \bar{\mathfrak{a}}$, so there is a bound for the rank of $f$:

$$\mathrm{rank}_{\mathbb{R}} df \leq 2 \dim_{\mathbb{C}} \mathfrak{a} \ ,$$

and the equality is attained if and only if $\mathfrak{a} \cap \bar{\mathfrak{a}} = \{0\}$, which is equivalent to the fact that $df(T_x M)$ has a complex structure, i.e. $\ker df_x \subset T_x M$ is $J$–invariant.

Since $\dim_{\mathbb{C}} \mathfrak{a} \leq 1$, we reach the following conclusion:

COROLLARY 6.7 (Sampson [109]). *Let $f\colon M \to N$ be a harmonic map from a compact Kähler manifold to a Riemannian manifold with universal cover $\tilde{N} = H_{\mathbb{R}}^n$. Then the rank of $f$ is at most two.*

Combining this statement with the Eells–Sampson existence theorem 5.8 we obtain topological restrictions on the maps from $M$ to $N$, for example:

COROLLARY 6.8. *Let $M, N$ be as above, and $f\colon M \to N$ a continuous map. Then the induced map on homology $f_*\colon H_k(M, \mathbb{Z}) \to H_k(N, \mathbb{Z})$ is zero if $k > 2$.*

**2.2. Hermitian symmetric spaces.** This is the extreme case where there are very large Abelian subalgebras.

Let $\tilde{N} = G/K$ be a Hermitian symmetric space. Fix an invariant complex structure on $\tilde{N}$, compatible with the Killing form. Then there is a decomposition
$$\mathfrak{p}^{\mathbb{C}} = \mathfrak{p}^{1,0} \oplus \mathfrak{p}^{0,1} \,.$$
The complex structure on $G/K$ fulfills the integrability condition
$$[\mathfrak{p}^{1,0}, \mathfrak{p}^{1,0}] \subset \mathfrak{p}^{1,0} \,.$$
Moreover, by the properties of the Cartan decomposition, $[\mathfrak{p}^{1,0}, \mathfrak{p}^{1,0}] \subset \mathfrak{k}^{\mathbb{C}}$, and $\mathfrak{p}^{1,0} \cap \mathfrak{k}^{\mathbb{C}} = \{0\}$, so we see that
$$[\mathfrak{p}^{1,0}, \mathfrak{p}^{1,0}] = 0 \,,$$
i.e., the entire subalgebra $\mathfrak{p}^{1,0}$ is Abelian.

EXAMPLE 6.9. Let us look at the case of $\tilde{N} = SU(p,q)/S(U(p) \times U(q))$, the parametrising space for complex $q$-planes in $\mathbb{C}^{p+q}$ for which the $(p,q)$ Hermitian form $|z_1|^2 + \cdots + |z_p|^2 - |z_{p+1}|^2 - \cdots - |z_{p+q}|^2$ is negative definite.

The real group $SU(p,q)$ has a complexification $SU(p,q)^{\mathbb{C}} \cong SL(p+q, \mathbb{C})$. The Cartan decomposition of the Lie algebra and corresponding decomposition of its complexification can be easily described in terms of block matrices as follows. First, $\mathfrak{su}(p,q)$ is the following subalgebra of $\mathfrak{sl}(p+q, \mathbb{C})$:
$$\mathfrak{su}(p,q) = \left\{ A = \begin{pmatrix} a & b \\ b^* & c \end{pmatrix} \mid a = -a^*, c = -c^* \right\} \,.$$
where $*$ denotes the conjugate transpose. In the Cartan decomposition $\mathfrak{k}$ corresponds to the diagonal blocks, and $\mathfrak{p}$ to the off-diagonal blocks. Consequently $\mathfrak{su}(p,q)^{\mathbb{C}} \cong \mathfrak{sl}(p+q, \mathbb{C})$, and
$$\mathfrak{k}^{\mathbb{C}} = \left\{ A \in \mathfrak{sl}(p+q, \mathbb{C}) \mid A = \begin{pmatrix} * & 0 \\ 0 & * \end{pmatrix} \right\}$$
$$\mathfrak{p}^{\mathbb{C}} = \left\{ A \in \mathfrak{sl}(p+q, \mathbb{C}) \mid A = \begin{pmatrix} 0 & * \\ * & 0 \end{pmatrix} \right\} \,.$$
The decomposition $\mathfrak{p}^{\mathbb{C}} = \mathfrak{p}^{1,0} \oplus \mathfrak{p}^{0,1}$ into eigenspaces of the complex structure corresponds simply to the decomposition
$$\mathfrak{p}^{1,0} = \left\{ A \in \mathfrak{p}^{\mathbb{C}} \mid A = \begin{pmatrix} 0 & * \\ 0 & 0 \end{pmatrix} \right\}$$
$$\mathfrak{p}^{0,1} = \left\{ A \in \mathfrak{p}^{\mathbb{C}} \mid A = \begin{pmatrix} 0 & 0 \\ * & 0 \end{pmatrix} \right\} \,.$$

One must note here that the spaces $\mathfrak{p}^{1,0}$ and $\mathfrak{p}^{0,1}$ are indeed conjugate to each other with respect to the relevant conjugation of $\mathfrak{sl}(p+q,\mathbb{C})$, namely the conjugation

$$\begin{pmatrix} a & b \\ c & d \end{pmatrix} \rightarrow \begin{pmatrix} -a^* & c^* \\ b^* & -d^* \end{pmatrix}$$

which leaves the real form $\mathfrak{su}(p,q)$ invariant (rather than the component-wise conjugation of matrices).

We thus see that $\mathfrak{p}^{1,0}$ is an Abelian subalgebra of $\mathfrak{p}^{\mathbb{C}}$ of complex dimension $pq = \dim_{\mathbb{C}} N$, and consisting entirely of nilpotent matrices.

In general, in the case of Hermitian locally symmetric spaces $\Gamma\backslash G/K$, although $\mathfrak{p}$ consists entirely of elements, its complexification $\mathfrak{p}^{\mathbb{C}}$ will have nilpotent elements. If an Abelian subalgebra $\mathfrak{a}$ is contained in $\mathfrak{p}^{\mathbb{C}}$ and consists entirely of semi-simple elements, then $\mathfrak{a}$ is a $K^{\mathbb{C}}$-conjugate to the complexification $\mathfrak{a}_0^{\mathbb{C}}$ of a real Abelian subalgebra $\mathfrak{a}_0 \subset \mathfrak{p}$. Consequently, by the definition of the rank of a symmetric space, $\dim \mathfrak{a} \leq \operatorname{rank}(G/K)$. But the last example shows that, in the presence of nilpotent elements, the maximum dimension of an Abelian subalgebra of $\mathfrak{p}^{\mathbb{C}}$ can be much larger than the rank of the symmetric space. The following theorem gives a first rough idea of the size of Abelian subalgebras of $\mathfrak{p}^{\mathbb{C}}$:

THEOREM 6.10 (Carlson–Toledo [**25**]). *Let $\tilde{N} = G/K$ be a symmetric space of non-compact type with no $SL(2,\mathbb{R})/SO(2)$-factor, and let $\mathfrak{a} \subset \mathfrak{p}^{\mathbb{C}}$ an Abelian subalgebra. Then:*
  (i) $\dim_{\mathbb{C}} \mathfrak{a} \leq \frac{1}{2} \dim_{\mathbb{C}} \mathfrak{p}^{\mathbb{C}}$.
  (ii) *If $\dim_{\mathbb{C}} \mathfrak{a} = \frac{1}{2} \dim_{\mathbb{C}} \mathfrak{p}^{\mathbb{C}}$, then $G/K$ is Hermitian symmetric, and $\mathfrak{a} = \mathfrak{p}^{1,0}$ for some invariant complex structure on $G/K$.*

PROOF. The proof consists of two steps.
*Step 1* is the following technical lemma, which is Proposition (4.1) of [**25**], to which we refer for the proof:

LEMMA 6.11. *If $\mathfrak{a}$ has a nonzero semi-simple element, then $\dim \mathfrak{a} < \frac{1}{2} \dim \mathfrak{p}^{\mathbb{C}}$.*

This is the only place in the proof of the theorem where the hypothesis that there is no $SL(2,\mathbb{R})/SO(2)$-factor is used.
*Step 2:* By Lemma 6.11, we can assume that $\mathfrak{a}$ consists entirely of nilpotent elements. Therefore

$$\mathfrak{a} \cap \bar{\mathfrak{a}} = \{0\},$$

as real elements are semi-simple, and $\mathfrak{a}$ is an isotropic space for the Killing form. Consequently,

$$\dim_{\mathbb{C}} \mathfrak{a} < \frac{1}{2} \dim_{\mathbb{C}} \mathfrak{p}^{\mathbb{C}}$$

and if there is equality of dimensions, then $\mathfrak{p}^{\mathbb{C}} = \mathfrak{a} \oplus \bar{\mathfrak{a}}$. This means that $\mathfrak{a}$ is $\mathfrak{p}^{1,0}$ for some complex structure which is compatible with the Killing form.

It remains to prove that this complex structure is invariant, i.e. that $[\mathfrak{k}, \mathfrak{a}] \subset \mathfrak{a}$. By the properties of the Cartan decomposition we have that $[\mathfrak{p}, \mathfrak{p}] = \mathfrak{k}$, thus

$$[\mathfrak{a} \oplus \bar{\mathfrak{a}}, \mathfrak{a} \oplus \bar{\mathfrak{a}}] = [\mathfrak{a}, \bar{\mathfrak{a}}] = \mathfrak{k}^{\mathbb{C}}$$

and since $\langle [[\mathfrak{a}, \bar{\mathfrak{a}}], \mathfrak{a}], \mathfrak{a} \rangle = \langle [\mathfrak{a}, \bar{\mathfrak{a}}], [\mathfrak{a}, \mathfrak{a}] \rangle = 0$, we have that

$$[[\mathfrak{a}, \bar{\mathfrak{a}}], \mathfrak{a}] \subset \mathfrak{a}^{\perp}.$$

Finally, since $\mathfrak{a}$ is maximal isotropic for the Killing form, we have $\mathfrak{a}^\perp = \mathfrak{a}$, thus $[\mathfrak{k}, \mathfrak{a}] \subset \mathfrak{a}$ and the proof is complete. $\square$

Among the corollaries of Theorem 6.10 we first mention a generalisation of Sampson's result (Corollary 6.7):

COROLLARY 6.12. *Let $f\colon M \to N$ be a harmonic map from a compact Kähler manifold to a Riemannian locally symmetric space of non–compact type which is not locally Hermitian symmetric. Then the rank of $f$ is strictly smaller than the dimension of $N$.*

Before explaining consequences of this corollary, we state and prove Siu's rigidity theorem [**120**], which was the original source of the ideas presented here.

THEOREM 6.13 (Siu), (Harmonic version). *Let $M$ be a compact Kähler manifold, $N = \Gamma\backslash G/K$ be a locally Hermitian symmetric space with no real hyperbolic plane factor, and $f\colon M \to N$ a harmonic map with $\mathrm{rank}(f) = \dim N$. Then $f$ is holomorphic for some invariant complex structure on $G/K$.*

There is another version of this theorem:

THEOREM 6.14 (Siu), (Topological version). *Let $M, N$ be manifolds as above, and $f\colon M \to N$ a continuous map, such that the fundamental homology class of $N$ belongs to the image $f_*(H_*(M, \mathbb{Z}))$. Then $f$ is homotopic to a holomorphic map for some invariant complex structure on $G/K$.*

A corollary is Siu's strengthening of Mostow's rigidity theorem for compact locally symmetric Hermitian manifolds:

THEOREM 6.15 (Siu's rigidity theorem). *Let $M$ be a compact Kähler manifold, $N$ a compact locally symmetric Hermitian manifold, and $f\colon M \to N$ a homotopy equivalence. Then $f$ is homotopic to a biholomorphism for some invariant complex structure on $G/K$.*

If, in addition, we assume that $M$ is also locally symmetric, we get the conclusion of Mostow's rigidity theorem for compact locally Hermitian symmetric manifolds, by using the fact that biholomorphic maps are isometric for the Bergmann metric.

PROOF. We prove the harmonic version, the other versions follow readily from topology and the Eells–Sampson theorem.

By assumption, there exists a point $x \in M$ such that $df(T_x^{1,0}M)$ has dimension $\frac{1}{2}\dim \mathfrak{p}^\mathbb{C}$. Therefore $df(T_x^{1,0}M) \subset \mathfrak{p}^{1,0}$ for an invariant complex structure on $N$, and $f$ satisfies the Cauchy-Riemann equations at $x$ for that complex structure.

The set $U = \{x \in M \mid \dim df(T^{1,0}M) = \frac{1}{2}\dim \mathfrak{p}^\mathbb{C}\}$ is nonempty by assumption, and as $f$ is harmonic, $d'f$ is then holomorphic in the complement of a proper subvariety. Hence $U$ is connected and dense, and so $f$ satisfies the Cauchy–Riemann equations with respect to that complex structure everywhere. $\square$

REMARK 6.16. For each Hermitian symmetric space target, Theorem 6.13 and its topological consequences can be strengthened in the following way. One need not assume that $\mathrm{rank}(f) = \dim N$, but only that $\mathrm{rank}(f)$ is sufficiently large, where "sufficiently large" is a specific function of the target, namely, larger than twice the maximum dimension of an Abelian subalgebra of $\mathfrak{p}^\mathbb{C}$ which is not contained in $\mathfrak{p}^{(1,0)}$. The same argument shows that harmonic maps of this sufficiently large rank must satisfy the Cauchy–Riemann equations. This maximum dimension of Abelian subalgebras not in $\mathfrak{p}^{(1,0)}$ has been calculated by Siu, see Theorem (6.7) of [121]. The analogous computation for symmetric targets which are not Hermitian symmetric, namely the maximum dimension of Abelian subalgebras of $\mathfrak{p}^{\mathbb{C}}$, and the subsequent refinement of Theorem 6.10, has been carried out for all symmetric spaces of classical type in [26], and in one exceptional case in [24].

## 3. Period domains

We begin with some examples of non–Kähler complex manifolds arising from geometric constructions[1].

Let $X = SO(2p,q)/(SO(2p) \times SO(q))$. This symmetric space is the parametrising space of $2p$–planes $A \subset \mathbb{R}^{2p+q}$ on which the quadratic form $x_1^2 + \cdots + x_{2p}^2 - y_1^2 - \cdots - y_q^2$ is strictly positive, henceforth called *positive planes*. This space is Hermitian symmetric if and only if $p = 1$ or $q = 2$

Define now $D = SO(2p,q)/(U(p) \times SO(q))$. This is a homogeneous complex manifold, with a complex structure given by the inclusion $U(1) \subset U(p)$ and the resulting action by conjugation on the tangent space. It parametrises positive $2p$–planes of $\mathbb{R}^{2p+q}$ with an almost complex structure, $\{A, J\}$, or equivalently the pairs $\{A \oplus A^\perp, J\}$.

There is a natural projection $\pi \colon D \to X$, which is a topologically trivial fibration with fiber $SO(2p)/U(p)$, the set of almost complex structures on $\mathbb{R}^{2p}$.

Due to the almost complex structure on $A$, the complexification of such a decomposition $\mathbb{R}^{2p+q} = A \oplus A^\perp$, $J \colon A \to A$ is a decomposition

$$\mathbb{C}^{2p+q} = (A \oplus A^\perp)_{\mathbb{C}} = A^{1,0} \oplus A^\perp_{\mathbb{C}} \oplus A^{0,1} \ .$$

Therefore, $D$ parametrises Hodge structures of weight 2, assigning to every such structure the decomposition with $A^{1,0} = H^{2,0}$, $A^{0,1} = H^{0,2}$, $A^\perp_{\mathbb{C}} = H^{1,1}$. Our assertion is:

THEOREM 6.17. *Let $\Gamma$ be a cocompact lattice in $SO(2p,q)$, where $p > 1$ and $q > 2$. Then the compact complex manifold $N' = D/\Gamma$ is not homotopy equivalent to a compact Kähler manifold.*

REMARK 6.18. This contrasts with the situation of Hodge structures of weight 1, whose parametrising space is the Siegel upper half plane which admits a Kähler metric.

REMARK 6.19. The manifold $D$ carries an invariant *indefinite* Kähler metric, which is positive definite on the tangent vectors to the fibers and negative definite on their orthogonal complements. Thus the manifolds $N'$ are symplectic.

REMARK 6.20. It is not hard to see that the complex structure just defined on $N'$ is not Kähler. This follows from the known fact that the deformation space of a fiber of $\pi$ is not compact. Thus the point of the theorem is that no complex structure (within the homotopy type) can carry a Kähler metric. Also, it follows from the proof that the theorem is true for all the locally homogeneous complex manifolds considered in [57] for which the associated space is not Hermitian symmetric.

---

[1]This Section, and the next one, are based in part on a lecture of J. Carlson.

PROOF OF THEOREM 6.17. Let $N'$ be as above, and let $N = \Gamma\backslash X$, so that $N$ is locally symmetric and $\pi$ induces a fibration $\pi: N' \to N$ whose fiber is denoted by $F$. Let $\omega$ be the Chern class of the anti–canonical bundle of $N'$, and let $\Omega$ be a volume form on $N$. Since $F$ is a homogeneous rational variety, its anti–canonical bundle is positive, thus $\omega^r > 0$ on $F$, where $r$ is the dimension of $F$. By Fubini's theorem

$$\int_{N'} \pi^*\Omega \wedge \omega^r > 0 .$$

Therefore $\pi^*\Omega$ is a nonzero cohomology class, and by Poincaré duality there exists a cycle $C$ in $N'$ such that

$$\int_{\pi_*C} \Omega = \int_C \pi^*\Omega \neq 0 .$$

Suppose that there existed a compact Kähler manifold $M$ and a homotopy equivalence $f: M \to N'$. Then $g = \pi \circ f: M \to N$ and the fundamental class of $N$ would be in the image of $g_*$, contradicting Theorem 6.14. $\square$

## 4. The factorisation theorem

In this Section we prove a result which strengthens Sampson's theorem stating that harmonic maps from compact Kähler manifolds to real hyperbolic spaces have rank at most two.

THEOREM 6.21 (Sampson [108], Carlson–Toledo [25]). *Suppose $M$ is a compact Kähler manifold, $N = H_\mathbb{R}^n/\Gamma$ the quotient of real hyperbolic space by a co-compact lattice, and $f: M \to N$ a nonconstant harmonic map. Then one of the following two statements holds:*
(i) *The map $f$ has rank one and its image is a closed geodesic in $N$.*
(ii) *The map $f$ has rank two and admits a factorisation $f = \psi \circ \varphi$,*

$$M \xrightarrow{\varphi} S \xrightarrow{\psi} N ,$$

*where $S$ is a compact Riemann surface, $\varphi$ is holomorphic and $\psi$ is harmonic.*

We give two applications of the factorisation theorem to the study of Kähler groups.

THEOREM 6.22. *Let $\Gamma$ be a lattice in $SO(1,n)$, with $n > 2$. Then $\Gamma$ is not a Kähler group.*

PROOF. By passing to a subgroup of finite index we may assume that $\Gamma$ has no torsion. Suppose there exists a compact Kähler manifold $M$ such that $\pi_1(M) \cong \Gamma$. The quotient $N = H_\mathbb{R}^n/\Gamma$ is an Eilenberg–Mac Lane space $K(\Gamma, 1)$, so there exists a continuous map $f: M \longrightarrow N$ realising any isomorphism of fundamental groups. As $N$ is negatively curved, by the Eells–Sampson existence theorem 5.8 $f$ can be chosen to be harmonic if $N$ is compact. If $N$ is not compact, since a lattice is Zariski dense, the representation of the fundamental group is reductive, so Corlette's existence theorem 5.12 applies, and we can also choose $f$ to be harmonic.

By the factorisation theorem 6.21, $f$ decomposes as

$$M \xrightarrow{\varphi} S \xrightarrow{\psi} N ,$$

with $S$ either a circle $S^1$ or a compact Riemann surface. The map $\varphi$ induces an injective homomorphism $\phi_*: \Gamma \hookrightarrow \pi_1(S)$. Therefore, the case $S = S^1$ is not

possible, because it would imply that $\Gamma \cong \mathbb{Z}$. For the same reason, $\Gamma = \pi_1(M)$ acts effectively on the universal cover $\tilde{S}$ of $S$, and the quotient $\tilde{S}/\Gamma$ is another Eilenberg–Mac Lane space $K(\Gamma, 1)$. Thus, there are homology isomorphisms
$$H_*(N) \cong H_*(\tilde{S}/\Gamma) \cong H_*(\Gamma) .$$
In the cocompact case the space $N = H_{\mathbb{R}}^n/\Gamma$ has $H_n(N, \mathbb{Z}) \cong \mathbb{Z}$, and as $S$ is a surface, $H_k(\tilde{S}/\Gamma, \mathbb{Z}) = 0$ for $k > 2$. Therefore, we reach a contradiction unless $n \leq 2$.

In the non–cocompact case one uses the fact that the cohomological dimension of $\Gamma$ is $n - 1$, which leads us to the same contradiction as above, as long as $n > 3$. Finally, if $\Gamma$ is not cocompact and $n = 3$, one uses in addition the fact that $H^2(\Gamma, \mathbb{Z}\Gamma)$ is not finitely generated, contrary to the case of a closed Riemann surface. (Details on the cohomological facts just quoted to treat the non–cocompact case can be found in [**18**].) $\square$

Our second application is:

THEOREM 6.23. *Let $\Phi$ be a Kähler group admitting a representation in a torsion-free, cocompact lattice $\Gamma \subset SO(1, n)$, whose image is not cyclic. Then $\Phi$ is a fibered Kähler group. In particular, $\Phi$ has a nonzero second Betti number.*

PROOF. As in the proof of the previous theorem, we may take a compact Kähler manifold $M$ with $\pi_1(M) \cong \Phi$, and by the Eells–Sampson theorem there is a harmonic map $f \colon M \longrightarrow N = H_{\mathbb{R}}^n/\Gamma$. Since the image of the representation is not cyclic, the map must factorise through a surjective holomorphic map to a compact Riemann surface $S$. For the same reason, the genus of $S$ must be at least two, which proves that $\Phi$ is fibered. Since $H^*(S, \mathbb{C}) \hookrightarrow H^*(M, \mathbb{C})$ is injective and $S$ is a $K(\pi_1(S), 1)$, the last assertion follows. $\square$

PROOF OF THE FACTORISATION THEOREM 6.21. Recall the notation $\mathfrak{a}_x = df(T_x^{1,0} M)$. We saw that $\dim \mathfrak{a}_x = 1$, so there are two possible cases:
(i) The space $df(T_x M \otimes \mathbb{C}) = \mathfrak{a}_x + \bar{\mathfrak{a}}_x$ has dimension one. Then we immediately check that case (i) of the theorem holds.
(ii) $\mathfrak{a}_x + \bar{\mathfrak{a}}_x$ has dimension two. Then $\mathfrak{a}_x \cap \bar{\mathfrak{a}}_x = 0$, and we must show that case (ii) of the theorem holds.
*Set-up:* Let $y_0 \in N$ be a regular value of $f$. We choose a neighbourhood $U$ of $y_0$ in the rank two image of $f$, such that $f$ is a smoothly trivial fibration over it:

$$\begin{array}{ccc} f^{-1}(U) & \xrightarrow{\cong} & f^{-1}(y_0) \times U \\ & f \searrow \quad \swarrow \pi_2 & \\ & U & \end{array}$$

Take a connected component $C_{y_0}$ of $f^{-1}(y_0)$, let $V$ be the connected component of $f^{-1}(U)$ containing $C_{y_0}$, and $C_y = f^{-1}(y) \cap V$ for $y \in U$.
*Claim 1:* The tangent spaces $T_x C_y$ are invariant under $J$.

To check this, for any $X \in T_x C_y = \ker df_x$, take its type decomposition $X = X' + X''$, with $X' = X - iJX$. Then
$$df(X) = df(X') + df(X'') = 0 .$$
Since $df(X') \in \mathfrak{a}_x$, $df(X'') \in \bar{\mathfrak{a}}_x$, and $\mathfrak{a}_x \cap \bar{\mathfrak{a}}_x = 0$, we have that
$$0 = df(X') = df(X) - i \, df(JX) .$$

Hence $df(JX) = i\,df(X)$, and our claim.

It follows from Claim 1 that the level sets $C_y$ are complex submanifolds of $M$.
*Claim 2:* There is a complex manifold structure on $U$ such that the restriction $V \to U$ is holomorphic.

The complex structure on $U$ is defined by setting $T^{1,0}_y U = d'f(T^{1,0}_x M) = \mathfrak{a}_x$ for $x \in C_y$. As $\mathfrak{a}_x \cap \bar{\mathfrak{a}}_x = 0$, we have

$$T^{1,0}_y S \oplus T^{0,1}_y U = T_y U \otimes \mathbb{C}\,.$$

This is the only possible definition of a complex structure on $TU$ making $f$ holomorphic, yet we must check that it is consistent:
*Subclaim:* $\mathfrak{a}_x$ is constant as $x$ varies in $C_y$.

This is so because the $\mathfrak{a}_x$ are complex lines in $T_y U \cong \mathbb{C}^2$ which contain no real vectors as $\mathfrak{a}_x \cap \bar{\mathfrak{a}}_x = 0$. By intersecting with an affine line outside the origin, we can parametrise the set of such lines with a Riemann sphere minus its equator, which corresponds to lines containing real vectors. The parametrising map $w\colon C_y \to \mathbb{C}P^1 \setminus \mathbb{R}P^1$ sending $x$ to $\mathfrak{a}_x$ is holomorphic, and $C_y$ is compact and connected. Thus $w$ is constant, and hence our subclaim.

We can now construct the sought factorisation of $f$ by an analytic continuation argument. By Chow theory, there is a complete parametrising space $S$ for the cycles $C_y \subset M$, and a universal cycle $Z \subset S \times M$ such that

$$Z \cap \{s\} \times M = Z_s \sim C_{y_0}\,,$$

where $Z_s$ is the cycle in $M$ corresponding to the point $s$.

Let $p\colon Z \to S$, $q\colon Z \to M$ be the projections onto the first and second factors, respectively. The map $q\colon Z \to M$ is surjective, because $Z$ is compact and the image of $q$ contains the open set $V$. Moreover, the cycles $C_y$ are pairwise disjoint, so $S$ must have dimension one. It remains to prove that $q$ is an isomorphism, and that $f$ is constant on the fibers of $\varphi = p \circ q^{-1}\colon M \to S$.

Let us show that $q$ is finite. The cycles $Z_s$ are all equivalent, and they are disjoint over $U$, so $[Z_s] \cdot [Z_t] = 0$. Thus the only way that $Z_s \cap Z_t \neq \emptyset$ for $s \neq t$ is that both cycles are reducible, and the intersection consists of a union of their components. But $\dim S = 1$, so there are only finitely many reducible fibers, and thus $q$ is finite-to-one. We already know by its definition that it is generically one-to-one, so we conclude that $q$ is a biholomorphism between $Z$ and $M$.

We now show that $f$ is constant on the fibers of $\varphi = p \circ q^{-1}$. Let $W$ be the set of regular points of the restrictions of $f$ to the fibers of $\varphi$. It is an open and dense subset, both in $M$ and in every fiber. Let $E$ be the holomorphic bundle over $W$ of vectors of type $(1,0)$ tangent to the fibers of $f$.

The section $d'f \in \Gamma(M, Hom(T^{1,0}M, f^*T^{\mathbb{C}}N))$ is holomorphic, so by restriction the section $d'f \in \Gamma(W, Hom(E, f^*TN^{\mathbb{C}}))$ is holomorphic. But $df$ vanishes on $V$, equivalently, so does $d'f$. Hence $d'f \equiv 0$ as a section of $Hom(E, f^*TN^{\mathbb{C}})$, and $f$ has to be constant on the fibers of $\varphi$.

What we have seen so far shows that there exists a continuous map $\psi\colon S \to N$ such that $f = \psi \circ \varphi$. By the Eells–Sampson theorem there exists then a harmonic map $\psi^*\colon S \to N$ homotopic to $\psi$. The maps $\psi^* \circ \varphi$ and $f$ are homotopic and both harmonic. The target space $N$ is strictly negatively curved, so by Hartman's uniqueness theorem 5.4 the maps must be equal. Thus $\psi = \psi^*$ is harmonic, and our proof is complete. $\square$

**4.1. Related results.** To end this Section, we mention some recent applications of these techniques obtained taking other spaces as targets.

THEOREM 6.24. (L. Hernández [**68**]) *Let $N$ be a Riemannian manifold such that its sectional curvature verifies $R < 0$, and is strictly $\frac{1}{4}$–pinched, i.e. there are functions $c, \alpha \colon N \to \mathbb{R}$ with $\alpha_x > \frac{1}{4}$ such that for every $x \in N$*

$$c_x \leq R_x \leq \alpha_x c_x < 0 \; .$$

*Then $\pi_1(N)$ is not a Kähler group.*

REMARK 6.25. Examples of manifolds verifying the hypothesis of the previous theorem have been obtained by Gromov and Thurston as suitable branched covers of compact hyperbolic space forms of dimension $> 3$.

There are comparable results on the rank of harmonic maps from compact Kähler manifolds to complex, quaternionic and Cayley hyperbolic spaces (see [**25**], [**24**]). Moreover, in the case of Cayley hyperbolic spaces one also has the comparable result on fundamental groups [**24**].

The above results lead us naturally to the following conjecture:

CONJECTURE 6.26. *Let $G$ be a simple algebraic group, $K$ a maximal compact subgroup, and suppose that the symmetric space $G/K$ is not Hermitian symmetric. Let $\Gamma \subset G$ be a lattice. Then $\Gamma$ is not a Kähler group.*

As evidence supporting this conjecture, we have Theorem 6.22, Proposition 7.10 and the case of the exceptional group $F_4$, for which the conjecture has been established in [**24**].

## 5. Non–linear groups

We have already seen how harmonic maps give restrictions on linear Kähler groups. Similar restrictions can be obtained by non–Abelian Hodge theory, cf. Chapter 7. The methods in Chapter 4, based on $L^2$–cohomology, give some restrictions on Kähler groups without any assumption about linear representations. The purpose of this Section is to give a rough sketch of a new theory developed by Gromov and Schoen that also gives restrictions on Kähler groups without any assumptions of linearity. It is reasonable to expect that the Gromov–Schoen theory of harmonic maps to buildings [**62**] and the more recent theory of Korevaar and Schoen of harmonic maps to more general non–positively curved spaces [**82**] will have significant applications to the study of Kähler groups.

In this Section we only consider the Gromov–Schoen theory of harmonic maps to trees [**62**], and illustrate its usefulness in relation to a historically interesting example, namely the *Higman 4–group*. This was the subject of Serre's question [**113**] discussed in Section 2 of Chapter 1 of this book.

The **Higman 4 group** can be defined by the presentation:

$$H = \langle x_1, x_2, x_3, x_4 \mid x_2 x_1 x_2^{-1} = x_1^2, x_3 x_2 x_3^{-1} = x_2^2, x_4 x_3 x_4^{-1} = x_3^2, x_1 x_4 x_1^{-1} = x_4^2 \rangle \; ,$$

and has the following two remarkable properties:

(i) $H$ is infinite, and
(ii) $H$ has no proper subgroup of finite index.

The second property means that there exists no nontrivial homomorphism from $H$ to a finite group, thus there cannot be a non–trivial homomorphism from $H$ to a linear group. For a complete proof of both these properties the reader is referred to [**113**].

We sketch the proof that $H$ is infinite, as it will help to show that it cannot be Kähler. As building blocks, we take the four isomorphic groups

$$\Gamma_i = \langle x_i, y_i \mid y_i x_i y_i^{-1} = x_i^2 \rangle \ .$$

Each $\Gamma_i$ is an infinite group, isomorphic to the upper–triangular subgroup (Borel subgroup) of $GL(2, \mathbb{Z}[\frac{1}{2}])$ by the map

$$y_i \longmapsto \begin{pmatrix} 1 & 1 \\ 0 & 1 \end{pmatrix}$$

$$x_i \longmapsto \begin{pmatrix} 2 & 0 \\ 0 & 1 \end{pmatrix} ,$$

and each of the subgroups $\langle x_i \rangle$, $\langle y_i \rangle$ is infinite cyclic. Let $\Gamma_{12}$ be the amalgamated product over $\mathbb{Z}$ defined by:

$$\Gamma_{12} = \Gamma_1 \underset{\mathbb{Z}}{*} \Gamma_2 = \langle x_1, y_1, x_2, y_2 \mid y_1 x_1 y_1^{-1} = x_1^2, y_2 x_2 y_2^{-1} = x_2^2, y_1 = x_2 \rangle,$$

and define $\Gamma_{34}$ as the analogous amalgamated product of $\Gamma_3$ and $\Gamma_4$ over $\mathbb{Z}$. Then $\langle x_1, y_2 \rangle$ and $\langle x_3, y_4 \rangle$ are free subgroups of $\Gamma_{12}$ and $\Gamma_{34}$ respectively. Thus $H$ is an amalgamated product of $\Gamma_{12}$ and $\Gamma_{34}$ over a free group of rank 2, obtained by identifying $y_4 = x_1$ and $y_2 = x_3$. This amalgamated structure shows that $H$ is infinite, and also that it acts on a tree. Thus we can apply the following theorem of Gromov and Schoen, whose proof, to be sketched in the next Section, relies on the study of harmonic maps to trees.

THEOREM 6.27 (Gromov–Schoen [**62**]). *Let $X$ be a compact Kähler manifold with fundamental group $\Gamma = \pi_1(X)$. Suppose that $\Gamma$ is an amalgamated product $\Gamma = \Gamma_1 \underset{\Delta}{*} \Gamma_2$, where the index of $\Delta$ in $\Gamma_1$ is $\geq 2$, and the index of $\Delta$ in $\Gamma_2$ is $\geq 3$. Then there exists a representation $\rho \colon \pi_1(X) \to \mathrm{Aut}\,(\mathbb{D}^2) \cong PSL(2, \mathbb{R})$, where $\mathbb{D}^2 = \{z \in \mathbb{C} \mid |z| < 1\}$ is the Poincaré disk, with discrete, cocompact image, and a holomorphic equivariant map from the universal cover of $X$:*

$$f \colon \tilde{X} \longrightarrow \mathbb{D}^2 \ .$$

REMARK 6.28. Even though the action of $\rho(\Gamma)$ on $\mathbb{D}^2$ may have fixed points, the $\Gamma$–equivariant map $f \colon \tilde{X} \to \mathbb{D}^2$ descends to a surjective holomorphic map $X \to \rho(\Gamma) \backslash \mathbb{D}^2$. Thus $X$ fibers over a curve, which may have genus zero. But $\Gamma$ has a subgroup of finite index which is fibered in the sense of Definition 2.12.

Before going into the proof, let us give two applications of this Theorem:

COROLLARY 6.29. *The Higman 4–group $H$ is not a Kähler group.*

PROOF. We have seen that $H$ is an amalgamated product $\Gamma' \underset{\Delta}{*} \Gamma'$, with $\Delta$ a subgroup of infinite index in both factors. Thus $H$ verifies the assumption of the theorem. On the other hand, $H$ has no nontrivial homomorphisms to linear groups, in particular to $PSL(2, \mathbb{R})$. □

This shows that $H$ is not the fundamental group of a smooth projective variety. As mentioned in subsection 2.3 of Chapter 1, every finitely presentable group is the fundamental group of some reducible variety. We still do not know whether $H$ can be the fundamental group of an irreducible, but singular, variety.

Our second application of Theorem 6.27 is to linear groups:

COROLLARY 6.30. *The group $\Gamma = SL(2, \mathbb{Z}[\frac{1}{p}])$ is not a Kähler group, for any prime $p$.*

PROOF. The group $\Gamma$ is an amalgamated product $\Gamma \cong SL(2, \mathbb{Z}) *_\Delta SL(2, \mathbb{Z})$, where $\Delta$ is the subgroup

$$\Delta = \left\{ \begin{pmatrix} a & b \\ c & d \end{pmatrix} \in SL(2, \mathbb{Z}) \mid c \equiv 0 \mod p \right\} \subset SL(2, \mathbb{Z}),$$

embedded by the identity in the first factor, and by

$$\begin{pmatrix} a & b \\ c & d \end{pmatrix} \mapsto \begin{pmatrix} a & pb \\ c/p & d \end{pmatrix}$$

in the second; see [**113**], II 1.4. The index of $\Delta$ in each factor is $p + 1 \geq 3$.

If $\Gamma$ were a Kähler group, then by Theorem 6.27 it would act properly discontinuously and cocompactly on the Poincaré disk, in particular it would have a subgroup of finite index with infinite Abelianisation. On the other hand, since $SL(2, \mathbb{Z}[\frac{1}{p}])$ is an irreducible lattice in $SL(2, \mathbb{Q}_p) \times SL(2, \mathbb{R})$, any finite index subgroup has finite Abelianisation, cf. [**90**], Chapter IV, Theorem 3.9. □

REMARK 6.31. The above argument applies to all $S$-arithmetic groups of rank $\geq 2$ which act freely on a tree.

REMARK 6.32. Observe that there is a great difference in the trees used to prove the last two Corollaries. The valence of a vertex of the tree associated to an amalgamated product decomposition is the index of the edge subgroup in the vertex subroup [**113**]. The vertices of the tree used for the proof of Corollary 6.30 have valence $p + 1$, so in particular the tree is locally compact. But the vertices of the tree used for the proof of Corollary 5 have infinite valence, thus the tree is not locally compact. This brings some technical and conceptual difficulties in the discussion of Step 1 below.

## 6. Harmonic maps to trees

Bass and Serre characterised amalgamated products of groups in terms of group actions on trees (see [**113**]). In particular, if $\pi_1(M)$ splits as a non-trivial amalgamated product, then there exists a suitable tree $X$ and a representation

$$\rho \colon \pi_1(M) \longrightarrow \mathrm{Aut}\,(X)\,.$$

Moreover, given such a representation one can form the associated bundle over $M$ with fiber $X$. Since $X$ is contractible, this bundle has a section, equivalently, there exists a $\rho$-equivariant contiuous map

$$f \colon \tilde{M} \longrightarrow X$$

from the universal cover of $M$ to $X$. We will define the concept of harmonicity in this context, and briefly discuss the existence of harmonic and pluriharmonic maps from Kähler manifolds to trees, paralleling the Siu–Sampson Theorem 6.1.

Let $M$ be a Riemannian manifold, and $(X,d)$ a metric space. We introduce, following Korevaar and Schoen [82], the notions of *energy* and *energy density* of $L^2$–maps $f\colon M \to X$. Denote by $S(TM) \subset TM$ the unit sphere bundle in the tangent bundle of $M$. For any $\varepsilon > 0$, define an approximate $\varepsilon$-energy density of $f$ as
$$e_\varepsilon(f)(x) = \int_{S(T_xM)} \frac{d^2(f(x), f(\exp(\varepsilon v)))}{\varepsilon^2} dvol \ .$$
Next, define the $\varepsilon$-energy of $f$ to be the linear functional $E_\varepsilon(f)$ on $C_c(M)$ (continuous functions with compact support) which assigns to $\phi \in C_c(M)$ the number
$$E_\varepsilon(f)(\phi) = \int_M \phi(x) e_\varepsilon(f)(x) dvol_M = \int_{S(T_*M)} \phi(x) \frac{d^2(f(x), f(\exp(\varepsilon v_x)))}{\varepsilon^2} dvol \ .$$
Observe that if $X$ is a Riemannian manifold and $f$ is smooth, one has that
$$\lim_{\varepsilon \to 0} e_\varepsilon(f)(x) = e(f)(x) \ ,$$
where $e(f)$ is the usual energy. If $M$ is compact, the usual total energy $E(f)$ of $f$ is given by either of the formulae
$$E(f) = \int_M e(f)(x) dvol \ = \lim_{\varepsilon \to 0} E_\varepsilon(f)(1) \ .$$
Korevaar and Schoen define $f$ to be of *finite energy* if
$$\sup_{\phi \in C_c(M),\ 0 \le \phi \le 1} (\limsup_{\varepsilon \to 0} E_\varepsilon(f)(\phi)) < \infty \ ,$$
and they prove the following theorem:

PROPOSITION 6.33 (Korevaar–Schoen [82]). *If $f$ has finite energy, then $\lim_{\varepsilon \to 0} e_\varepsilon(f)$ exists as a measure on $M$.*

Therefore, under the hypothesis of this Proposition, one can make the following definitions:

DEFINITION 6.34. (i) The *energy density* of $f$ is the measure $e(f)$.
(ii) The *energy* of $f$ is the integral $E(f) = \int_M e(f)$.
(iii) A map $f\colon M \to X$ is *harmonic* if it is locally energy minimising, i.e., it minimises energy when restricted to all sufficiently small domains $\Omega \subset M$ with smooth boundary and when compared with all maps with the same boundary values.
(iv) Let $M$ be compact Kähler. A map $f\colon M \to X$ to a metric space $X$ is *pluriharmonic* if its restrictions to all germs of holomorphic curves in $M$ are harmonic.

EXAMPLE 6.35. Set $M = \mathbb{R}^2$, and $X = \{(x,y) \in \mathbb{R}^2 \mid x = \pm y\}$ the space formed by the two diagonals of the plane, with the path–length metric. The product function $(x,y) \mapsto xy$ defines a foliation in $\mathbb{R}^2$, such that all its leaves, either smooth or singular, cut $X$ in exactly one point. This defines a map
$$\begin{aligned} f : M = \mathbb{R}^2 &\longrightarrow X \\ (x,y) &\longmapsto (x \cdot |y|, |x| \cdot y) \ , \end{aligned}$$
which is harmonic. To check this, consider the projection
$$\pi : X \to X/\{\pm \mathrm{Id}\} \cong \mathbb{R} \ .$$

The identification $X/\{\pm \mathrm{Id}\} \cong \mathbb{R}$ is a homeomorphism, and an isometry in the path metrics. The projection $\pi$ is a 2-to-1 covering branched only at 0. It is easily checked now that the composition $\pi \circ f$ is a harmonic map in the classical sense, and harmonicity of $f$ follows from this.

In this example, it is important to note that $f^{-1}(0)$ consists of the union of the $x$– and $y$–axes, but the behaviour of $f$ at the origin is more complicated than at any other point of $f^{-1}(0)$. Namely, at any $(x_0, y_0)$ where $x_0 \neq 0$ or $y_0 \neq 0$, the point $(x_0, y_0)$ has a neighbourhood which is mapped by $f$ to a geodesic in $X$, thus to a space intrinsically isometric to $\mathbb{R}$, and the restriction of $f$ to this neighbourhood is an ordinary harmonic function. Such a point is called a *regular* point of $f$. A point where no such neighbourhood exists is called a *singular* point of $f$. This is the terminology used in [**62**] for maps to trees, and similar terminology (using flats rather than geodesics) for maps to buildings.

With this background on harmonic maps to trees, we can now very briefly describe the steps in the proof of Theorem 6.27. This is just a brief guide to the relevant sections of [**62**], which the reader will have to consult for details.

*Step 1. Existence of harmonic maps:* For finite trees this is a special case of Theorem 4.4 of [**62**], and the analogous result would hold for locally finite trees in the equivariant case. But the tree needed for the Higman group is not locally finite, so a more subtle existence theorem is needed. See the discussion in §9 of [**62**], on harmonic maps to $\mathbb{Z}$–trees. See also the general discussions in [**82**], where maps to non–locally compact spaces are treated and the behaviour of minimising sequences is analyzed. We will assume existence for the rest of this discussion.

*Step 2. Harmonic maps are pluriharmonic:* This requires the extension of the Siu–Sampson Theorem 6.1 to maps to trees.

THEOREM 6.36 ([**62**]). *Let $M$ be a compact Kähler manifold and $X$ a tree. Then harmonic maps $f : M \to X$ are pluriharmonic.*

This is a special case of Theorem 7.3 of [**62**], which in turn is based on their Theorem 6.4. We can describe the ideas involved roughly as follows. Suppose that the set $\Sigma$ of singular points of $f$, as defined above, formed a submanifold of codimension 2, as is the case in the above example. Then to carry through the proof we gave of Theorem 6.1 all we need to do is to prove the only global step in the argument, namely

$$\int_{M\setminus \Sigma} d(\langle d_\nabla d^c f \wedge d^c f \rangle \wedge \omega^{n-2}) = 0 .$$

Thus, if $S_\varepsilon(\Sigma)$ denotes the boundary of the $\varepsilon$–tubular neighbourhood of $\Sigma$, this is equivalent to

$$lim_{\varepsilon \to 0} \int_{S_\varepsilon(\Sigma)} \langle d_\nabla d^c f \wedge d^c \rangle \wedge \omega^{n-2} = 0 .$$

Thus one needs an estimate on the growth of the second derivatives of $f$ in a neighbourhood of $\Sigma$ as $\varepsilon \to 0$. Theorem 6.4 of [**62**] asserts that $\Sigma$ has Hausdorff codimension 2 in $M$ and gives, in a more precise form, the required estimate on second derivatives. From this their Theorem 7.3, pluriharmonicity of harmonic maps, follows easily, as indicated.

*Step 3. The factorisation theorem.* By this we mean the derivation of Theorem 6.27 from the pluriharmonicity of the map to a tree. This is a factorisation theorem, in the spirit of factorisation theorems, of various degrees of subtlety and difficulty, that we have encountered before, namely Theorem 2.7, Theorem 4.14, Scholium 4.18, Theorem 4.28, and Theorem 6.21. The situation here is perhaps closest to the situation of Chapter 4 in that the harmonic map gives a singular foliation of $M$ by possibly non–compact complex subvarieties of codimension one, namely the foliation of $M$ by maximal complex subvarietes of the fibers of $f$. Note that the foliations of Chapter 4 arise in the same way, as maximal complex subvarieties of the fibers of a harmonic function. One added difficulty here is that the foliation is not given by a one–form, but is instead given by a quadratic differential. The issue, as in Chapter 4 is to prove that the leaves of this foliation are all compact. The details are subtle, and are treated in §9 of [**62**], and the same subtleties are also treated by Simpson in [**117**].

CHAPTER 7

# Non–Abelian Hodge theory

## 1. Basic concepts

Classical Hodge theory gives a description of the complex cohomology of compact Kähler manifolds. This description is based on the identification of the de Rham cohomology groups of a compact Riemannian manifold $X$ with the linear space of harmonic $k$–forms,

$$H^k_{\mathrm{dR}}(X, \mathbb{C}) \cong \{\omega \in \mathcal{E}^*(X) \mid \Delta\omega = dd^*\omega + d^*d\omega = 0\} \,.$$

When $X$ is moreover Kähler, there is an analogous identification of the Dolbeault cohomology groups with the space of $\Delta''$–harmonic forms:

$$H^q_{\mathrm{Dol}}(X, \Omega^p) \cong \{\omega \in \mathcal{E}^{p,q}(X) \mid \Delta''\omega = \bar{\partial}\bar{\partial}^*\omega + \bar{\partial}^*\bar{\partial}\omega = 0\} \,.$$

This and the equality of Laplacians

$$\Delta = 2\Delta''$$

imply the Hodge decomposition of the complex cohomology of compact Kähler manifolds:

$$H^k_{\mathrm{dR}}(X, \mathbb{C}) = \bigoplus_{p+q=k} H^q_{\mathrm{Dol}}(X, \Omega^p) \,.$$

This decomposition may be interpreted as an action of the real algebraic group $\mathbb{C}^*$ on the cohomology groups $H^k_{\mathrm{dR}}(X, \mathbb{C})$, given by

$$z \circ \omega = z^p \bar{z}^q \omega \,,$$

for $z \in \mathbb{C}^*$ and $\omega \in H^q(X, \Omega^p)$. This action preserves the real cohomology subspace $H^k(X, \mathbb{R}) \subset H^k(X, \mathbb{C})$.

The non–Abelian analogues of Hodge structures that we will examine are given by actions of $\mathbb{C}^*$ on moduli spaces of flat bundles.

**1.1. Harmonic flat bundles, Higgs bundles and variations of Hodge structure.** Given a manifold $X$ and an arbitrary group $G$, the first cohomology set of $X$ with coefficients in $G$ is

$$H^1(X, G) = Hom\,(\pi_1(X), G)/G$$

where the quotient on the right–hand–side is by the action of $G$ by conjugation. This space parametrises isomorphism classes of $G$–local systems over $X$.

To begin with, we take $G = GL(k, \mathbb{R})$; there is an analogous story in the complex case. Linear local systems are equivalently characterised as flat vector bundles, so in this case $H^1(X, G)$ is the set of isomorphism classes of real rank $k$ vector bundles with a flat connection $(E, \nabla)$. There is a concept of harmonicity in this context.

DEFINITION 7.1. A *harmonic flat bundle* is a triple $(E, \nabla, h)$ formed by a flat bundle $(E, \nabla)$ and a metric $h$ on $E$ which is harmonic, i.e., the map between Riemannian manifolds

$$h \colon \tilde{X} \longrightarrow GL(k, \mathbb{R})/O(k)$$

induced by the pullback of $E \to X$ to the universal cover $\tilde{X}$ of $X$ is harmonic.

A flat vector bundle is said to be *reductive* if every invariant subbundle with respect to the connection has an invariant complementary subbundle. This is equivalent to the condition that the Zariski closure of its monodromy be a reductive group. The analogue of the Hodge theorem in this context is the following theorem, obtained by combining Theorems 5.12 and 5.16.

THEOREM 7.2. *An isomorphism class of flat bundles $[(E, \nabla)] \in H^1(X, G)$ has a harmonic representative if and only if $(E, \nabla)$ is reductive. This representative is unique up to connection–preserving isomorphism.*

In view of this theorem we will center our attention on the isomorphism classes of reductive bundles.

DEFINITION 7.3. The *Betti moduli space* of reductive bundles is the quotient

$$\mathcal{M}_{\mathrm{B}} = Hom^{red}(\pi_1(X), G)/G$$

of the space of reductive representations of $\pi_1(X)$ in $G = GL(k, \mathbb{R})$ by the action of $G$ given by conjugation.

The Betti moduli space $\mathcal{M}_{\mathrm{B}}$ is a real affine variety because it coincides with the quotient of the affine variety $Hom(\pi_1(X), G)$ by the action of $G$ in the sense of Geometric Invariant Theory. There is a continuous surjective map

$$H^1(X, G) \longrightarrow \mathcal{M}_{\mathrm{B}},$$

because the orbit of any representation always has a unique reductive homomorphism in its closure.

Given a vector bundle with flat connection $(E, \nabla)$, and an arbitrary metric $h$ on it, there is a unique decomposition of the connection as

$$\nabla = D + \theta,$$

where $D$ is a connection compatible with the metric, and $\theta$ is a 1–form with values in the symmetric endomorphisms of $E$. Decomposing the flatness condition $\nabla^2 = 0$ into symmetric and skew–symmetric components, we have respectively

$$D^2 + \frac{1}{2}[\theta, \theta] = 0,$$
$$D\theta = 0.$$

The 1–form $\theta$ turns out to be the differential of the metric–induced map $h \colon \tilde{X} \to GL(k, \mathbb{R})/O(k)$, while $D$ is essentially the pullback of the Levi–Civita connection, so the condition for harmonicity of the bundle is the equation

$$D^*\theta = 0.$$

The analogue of the Laplacian identity $\Delta = 2\Delta''$ is the following version of the Siu–Sampson Bochner formula (compare Chapter 6).

THEOREM 7.4 (Siu–Sampson). *If $X$ is compact Kähler and $(E, \nabla, h)$ is a harmonic flat bundle, then:*
(i) $(D'')^2 = 0$, *i.e., the (0,1)-component $D''$ of the complexified Levi–Civita connection $D$ induces a holomorphic structure on $E_{\mathbb{C}} = E \otimes \mathbb{C}$,*
(ii) $[\theta^{1,0}, \theta^{1,0}] = 0$,
(iii) $D''\theta^{1,0} = 0$, *i.e., $h\colon \tilde{X} \to GL(k,\mathbb{R})/O(k)$ is a pluriharmonic map.*

This result motivates the next definition.

DEFINITION 7.5 (Hitchin, Simpson). A (real) *Higgs bundle* over $X$ is a pair $(E, \phi)$, where
(i) $E$ is a holomorphic vector bundle with a nondegenerate holomorphic symmetric bilinear form, and
(ii) $\phi \in H^0(X, \Omega^1_X \otimes \mathrm{Sym}\,(E))$, satisfying

$$[\phi, \phi] = 0 \;,$$

where $\mathrm{Sym}\,(E)$ is the bundle of symmetric endomorphisms of $E$. (Note: The bundle $E$ which occurs here is the complexification of the one which occurs above.) For a *complex Higgs bundle*, we drop the condition on the existence of a symmetric bilinear form on $E$, and $\phi$ is merely required to be a holomorphic $End(E)$-valued 1-form which commutes with itself.

Given a harmonic bundle $(E, \nabla, h)$, one can define an associated one-parameter family of complex flat bundles by setting

$$\nabla_\lambda = D + \lambda \theta^{1,0} + \lambda^{-1} \theta^{0,1} \;.$$

One can check that the flatness condition $\nabla^2_\lambda = 0$ holds if conditions (i)-(iii) of Theorem 7.4 hold, and if $|\lambda| = 1$, then $(E, \nabla_\lambda, h)$ is again a harmonic bundle. In this way we obtain an action of $U(1)$ on the Betti moduli space $\mathcal{M}_B$. We will see that the study of the fixed points of this action of $U(1)$ on $\mathcal{M}_B$ is very relevant to the study of representations of Kähler groups.

If $(E, \nabla, h)$ represents a fixed point of the $U(1)$-action, then there is an isomorphism $(E, \nabla, h) \cong (E, \nabla_\lambda, h)$, i.e., there exists a gauge transformation $g_\lambda \colon E \to E$ such that:
(i) $g_\lambda$ preserves the metric $h$ and the connection $D$, and
(ii) $g^*_\lambda \theta^{1,0} = \lambda \theta^{1,0}$.
The choice of $g_\lambda$ is ambiguous up to the elements of the group of gauge transformations that preserve all of the structure. Assume that the fixed point $[(E, \nabla)]$ is irreducible. Then the group of automorphisms of $(E, \nabla, h)$ is $\{\pm 1\}$, and therefore there is an extension of groups

$$1 \longrightarrow \{\pm 1\} \longrightarrow K \longrightarrow U(1) \longrightarrow 1$$

with $K \subset \mathrm{Aut}\,(E, D, h)$. Thus we may find a homomorphism

$$\alpha \colon U(1) \longrightarrow \mathrm{Aut}\,(E, D, h)$$

such that $\alpha(\lambda) = g_{\lambda^N}$, for $N = 1$ or $2$. This homomorphism defines a decomposition of the complexified bundle $E_{\mathbb{C}}$ into eigenbundles:

$$E_{\mathbb{C}} = \bigoplus_{p \in \frac{1}{N}\mathbb{Z}} E^{p,-p} \;.$$

The group $\mathrm{Aut}\,(E,D,h)$ preserves $D''$, so this decomposition is holomorphic with respect to $D''$. Moreover, the fact that $g_\lambda^* \theta^{1,0} = \lambda \theta^{1,0}$ implies that

$$\theta^{1,0}(E^{p,-p}) \subset \Omega_X^1 \otimes E^{p-1,1-p}\,.$$

Now set
$$\sigma = g_{-1} = \alpha(e^{\frac{\pi i}{N}})$$
and
$$\varepsilon = (g_{-1})^2 = \pm 1\,.$$

Define a bilinear form $H$ on $E$ by $H(\cdot,\cdot) = h(\cdot,\sigma\cdot)$.

*Claim 1:* The bilinear form $H$ is parallel with respect to the flat connection, i.e., $\nabla H = 0$.

We compute:
$$(\nabla H)(v,w) = d(H(v,w)) - H(\nabla v, w) - H(v, \nabla w)\,.$$

The first term on the right–hand–side can be computed as
$$d(H(v,w)) = d(h(v,\sigma w)) = h(\nabla v, \sigma w) + h(v, \nabla_{-1}\sigma w)$$
$$= H(\nabla v, w) + h(v, \sigma \nabla w) = H(\nabla v, w) + H(v, \nabla w)\,.$$

*Claim 2:* The bilinear form $H$ is $\varepsilon$–symmetric.

This can be easily computed:
$$H(s_1, s_2) = h(s_1, \sigma s_2) = h(\sigma s_2, s_1) = h(\sigma^2 s_2, \sigma s_1) = \varepsilon H(s_2, s_1)\,.$$

Thus, we have sketched the proof of the following result:

THEOREM 7.6 (Simpson [116]). *Any fixed point of the $U(1)$–action on the Betti moduli space $\mathcal{M}_B$ arises from a real variation of Hodge structure (see below for the definition).*

Let us recall the concept of a real variation of Hodge structure:

DEFINITION 7.7. A *real variation of Hodge structure* ($\mathbb{R}$-VHS) of weight $k$ is the following set of data:
(i) A real vector bundle $E$ together with a smooth decomposition of its complexification into complex subbundles
$$E_\mathbb{C} = \bigoplus_{p,q \geq 0, p+q = k} E^{p,q}$$
such that $\overline{E^{p,q}} = E^{q,p}$,
(ii) a flat connection $\nabla$ on $E$,
(iii) a $(-1)^k$–symmetric parallel nondegenerate bilinear form $H\colon E \times E \to \mathbb{R}$, such that:
(1) the subbundles $F^r = \bigoplus_{p \geq r} E^{p,q}$ are $\nabla''$–holomorphic,
(2) (Griffiths transversality) the sheaves $\mathcal{F}^r$ of holomorphic sections of $F^r$ satisfy:
$$\nabla' \mathcal{F}^r \subset \Omega_X^1 \otimes \mathcal{F}^{r-1}\,,$$
(3) Denoting by $C$ the operator given by multiplication by $i^{p-q}$ in $E^{p,q}$, the bilinear form $H(\cdot, C\cdot)$ is positive definite.

More generally, if we drop the condition that the bundle be real, then we find that a fixed point of the $U(1)$–action gives rise to what is known as a complex variation of Hodge structure.

## 1. BASIC CONCEPTS

**1.2. Groups of Hodge type and the volume form.** The existence of a $(-1)^k$–symmetric bilinear form $H$ on an oriented flat bundle $(E, \nabla)$ implies that the monodromy is a representation $\rho\colon \pi_1(X) \to G = Sp(2n, \mathbb{R})$ or $SO(m,n)$, depending on the parity of $k$. Fixing the dimensions of subbundles $E^{p,q}$ in a decomposition of $E \otimes \mathbb{C}$ as in 7.7 determines a compact subgroup $M \subset K \subset G$ such that $G/M$ is a complex manifold. Complex manifolds arising in this way are known as period domains (compare Section 3 in Chapter 6).

PROPOSITION 7.8. *A real variation of Hodge structure over $X$ is equivalent to a flat $G$-vector bundle, with $G = Sp(2n,\mathbb{R})$ or $SO(m,n)$ according to weight, and an equivariant holomorphic map (the period map)*

$$\tilde{X} \longrightarrow G/M \ .$$

There is an underlying harmonic bundle, obtained by composing the period map with the projection $G/M \to G/K$. In particular, all flat bundles arising from variations of Hodge structure are reductive.

DEFINITION 7.9. A reductive Lie group is of *Hodge type* if it has a compact Cartan subgroup. The groups of Hodge type are those which act as automorphism groups of period domains; they may also be characterised as the groups which have a discrete series.

PROPOSITION 7.10 (Simpson [**116**]). *If $(E, \nabla, H)$ is a $\mathbb{R}$-VHS, then the real Zariski closure of the monodromy is a group of Hodge type.*

A consequence of this for Kähler groups is:

COROLLARY 7.11 ([**116**]). *If $G$ is a simple Lie group not of Hodge type, and $\Gamma$ is a lattice in $G$, then $\Gamma$ is not the fundamental group of a compact Kähler manifold.*

To prove the Corollary, we can assume that the real rank of $G$ is at least 2, since the only groups of rank 1 but not of Hodge type are those locally isomorphic to $SO(2n+1, 1)$, and these are covered by a result of Carlson–Toledo discussed in Chapter 6. In the higher rank case, we can appeal to results of Weil (in the case where $\Gamma$ is cocompact) and Margulis to conclude that $\Gamma$ is locally rigid. If it is also a Kähler group, then Proposition 2 leads to a contradiction.

PROOF OF PROPOSITION 7.10. Let $G$ be the Zariski closure of the monodromy of $(E, \nabla)$, $P$ its associated principal $G$-bundle, and $V = Ad(P) \subset End(E)$ the composition of some representation $\rho\colon \pi_1(X) \to G$ inducing $P$ with the adjoint representation of $G$.

The polarisation $H$ of the $\mathbb{R}$-VHS induces an inner product $\langle \cdot, \cdot \rangle$ on $V$. Let

$$\tau: V_\mathbb{C} \to V_\mathbb{C}$$

be the complex conjugation, and

$$\sigma: V \to V$$

the gauge transformation given by $\alpha(e^{(\pi i)/N})$. Obviously $G = G_\mathbb{C}^\tau$, i.e., the real elements of the group are those fixed under conjugation. Set now $G^c = G_\mathbb{C}^{\sigma\tau}$. The form $\langle \cdot, \sigma\tau\cdot \rangle$ is positive definite Hermitian, so $G^c$ is a compact real form of $G$. The gauge transformation $\sigma$ belongs to $G^c$, so there is a maximal torus $T \subset G^c$ containing $\sigma$. As $T$ is fixed by $\sigma\tau$, $T$ is in $G_\mathbb{C}^\tau = G$ and it is a compact Cartan subgroup. Thus $G$ is of Hodge type. □

We will conclude this section with another application to the study of Kähler groups. Let $G = Iso(D)$, for $D$ an irreducible bounded symmetric domain.

DEFINITION 7.12. If $\rho\colon \pi_1(X) \to G$, then $vol(\rho)$ is the cohomology class of the pullback of the volume form of $D$ by any $\rho$-equivariant map $f\colon \tilde{X} \to D$.

For suitable $G$, if $\Gamma \subset G$ is a lattice then the inclusion is locally rigid among all homomorphisms of $\Gamma$ to $G$, i.e., it gives an isolated point in the moduli space of all homomorphisms. The following result implies that this property is to a certain extent hereditary under pullback by surjective holomorphic maps.

THEOREM 7.13 ([**37**]). *If $G$ is not $SU(k,1)$ or $SO(2k+1,2)$, and $X$ is compact Kähler, then $vol(\rho) \neq 0$ implies that the representation $\rho$ is locally rigid as a representation in the complexification of $G$.*

SKETCH OF PROOF. If $\rho$ is not locally rigid we can find a family of representations $\rho_t\colon \pi_1(X) \to G_{\mathbb{C}}$ such that:
1. $\rho_0 = \rho$, and
2. neither the images $\rho_t(\pi_1(X))$ nor any of their conjugates are contained in $G$.

For each $\rho_t$ we have a harmonic map $f_t\colon \tilde{X} \to G/K$, and for $t$ small this map is a deformation of the original harmonic map $f$ of $\rho$.

By the Siu–Sampson Theorem 7.4, $[\theta^{1,0}, \theta^{1,0}] = 0$. Therefore the image of the holomorphic tangent bundle $df(T_m^{1,0}X) \subset \mathfrak{g}_{\mathbb{C}}$ is fiberwise an Abelian subalgebra. For the initial representation, the corresponding map is holomorphic with generically maximal rank, so the subalgebra we obtain at a generic point is the algebra $\mathfrak{p}^{1,0}$ which determines the invariant complex structure on $G/K$. For the groups under consideration, any small deformation of $\mathfrak{p}^{1,0}$ as an Abelian subalgebra turns out to be conjugate to $\mathfrak{p}^{1,0}$. Furthermore, the 1–jets of the maps $f_t$ are generically 1–jets of holomorphic maps to some conjugate of $D \subset G_{\mathbb{C}}/H$. One can show that this together with the harmonicity of the $f_t$ implies that the representations $\rho_t$ are fixed points for the $U(1)$–action on the Betti moduli space. But this contradicts the fact that $\rho_t(\pi_1(X)) \not\subset G$ for $t \neq 0$, even after conjugation. Hence our statement. □

## 2. Yang–Mills equations and the $\mathbb{C}^*$–action on Higgs bundles

In the previous section, a $U(1)$–action on the Betti moduli space of reductive bundles $\mathcal{M}_B$ was described, by means of the harmonic representatives $(E, \nabla, h)$ of isomorphism classes of reductive flat bundles. Our next goal is to extend this action to a $\mathbb{C}^*$–action on $\mathcal{M}_B$. We will do this through the use of Higgs bundles, which will be taken to be complex from now on. First, we reformulate the definition using a certain differential operator $\mathcal{D}''$.

DEFINITION 7.14. Let $(E, \phi)$ be a Higgs bundle. The differential operator
$$\mathcal{D}''\colon \mathcal{E}^k(X, E) \to \mathcal{E}^{k+1}(X, E)$$
is defined as
$$\mathcal{D}'' = D''_E + \phi \,.$$

PROPOSITION 7.15. *Let $E$ be a holomorphic vector bundle and $\phi$ a 1–form with values in $End(E)$. The pair $(E, \phi)$ is a Higgs bundle if and only if*
$$(\mathcal{D}'')^2 = (D''_E + \phi)^2 = 0 \,.$$

## 2. YANG–MILLS EQUATIONS AND THE $\mathbb{C}^*$-ACTION ON HIGGS BUNDLES

In order to define a $\mathbb{C}^*$–action on the moduli space of flat bundles, given a Higgs bundle $(E, \phi)$ corresponding to some flat bundle, we seek flat bundles corresponding to $(E, \lambda\phi)$ for every $\lambda \in \mathbb{C}^*$. This amounts to finding the appropriate Hermitian metric on $E$, using a generalisation of the standard construction of a canonical connection on a holomorphic bundle with a Hermitian metric.

Let $h$ be a Hermitian metric on $E$. If $(E, \phi)$ is a Higgs bundle, then $(\overline{E^*}, \phi^*)$ is an anti–Higgs bundle, i.e.,

(i) $\overline{E^*}$ is an antiholomorphic bundle, and

(ii) the form $\phi^* \colon \overline{E^*} \to \overline{\Omega^1} \otimes \overline{E^*}$ is antiholomorphic and satisfies $[\phi^*, \phi^*] = 0$.

A Hermitian metric is equivalent to an isomorphism $h \colon E \to \overline{E^*}$. Thus we can pull back the operator $D'_{\overline{E}} + \phi^*$ to obtain an operator $\mathcal{D}'_h$, and therefore a connection on $E$:

$$\nabla_h = \mathcal{D}'_h + \mathcal{D}'' = D_h + \phi + \phi^* \, .$$

Here $D_h$ is the canonical Hermitian connection associated to the holomorphic structure and the metric. The point is to choose the metric $h$ appropriately to get a flat connection $\nabla_h$. Let us consider first a weakened version of this problem. Let $\Lambda$ denote contraction of a form with the Kähler form $\omega$.

*Problem:* Find a Hermitian metric $h$ such that

$$\Lambda(\nabla_h^2) = \Lambda F_h = \lambda Id \, .$$

This equation is the *Higgs–Hermitian–Yang–Mills equation* (henceforth abridged to HHYM).

REMARK 7.16. The scalar $\lambda$ is subject to topological constraints:

$$\lambda \operatorname{rk}(E) = \frac{\int_X \operatorname{Tr} F_h \wedge \omega^{n-1}}{vol(X)} = \frac{\int_X c_1(E) \wedge \omega^{n-1}}{vol(X)} \, .$$

REMARK 7.17. If the curvature satisfies $\Lambda F_h = 0$, then $F_h$ is a primitive $(1, 1)$–form. Such forms are orthogonal to the Kähler form, and are anti–self–dual, meaning

$$*F_h = -\frac{1}{(n-2)!} F_h \wedge \omega^{n-2} \, .$$

(Compare Section 1 of Chapter 6.) Then

$$\int_X |F_h|^2 = -\frac{1}{(n-2)!} \int \operatorname{Tr}(F_h \wedge F_h) \wedge \omega^{n-2} = \frac{1}{2(n-2)!} \int_X c_2(E) \wedge \omega^{n-2}$$

Thus, if the HHYM equation has a solution $h$, and $c_1(E) = c_2(E) = 0$, then $\nabla_h$ is a flat linear connection.

### 2.1. Solutions of the HHYM equation.

DEFINITION 7.18. (i) A *Higgs subsheaf* of $(E, \phi)$ is a holomorphic subsheaf $\mathcal{F}$ of the associated sheaf of sections $\mathcal{E}$, with the property

$$\phi(\mathcal{F}) \subset \Omega^1 \otimes \mathcal{F} \, .$$

(ii) A Higgs bundle $(E, \phi)$ is *stable* if for every nontrivial saturated Higgs subsheaf $(F, \phi_F)$ the following inequality holds:

$$\mu(F) = \frac{\int_X c_1(F) \wedge \omega^{k-1}}{\operatorname{rk}(F)} < \frac{\int_X c_1(E) \wedge \omega^{k-1}}{\operatorname{rk}(E)} = \mu(E) \, .$$

Stability turns out to be the necessary and sufficient condition for existence of solutions to the HHYM equation.

PROPOSITION 7.19. *Suppose that the Higgs bundle $(E, \phi)$ admits a solution to the HHYM equation and is irreducible. Then $(E, \phi)$ is stable.*

PROOF. (Sketch) The Proposition is a consequence of two principles which may be loosely stated as:

1. curvature decreases in Higgs subbundles, and
2. curvature is evenly distributed over $E$.

Suppose that $F \subset E$ is a Higgs subbundle. The choice of a Hermitian metric $h$ produces a vector bundle decomposition $E = F \oplus F^\perp$, which induces decompositions

$$D_E'' = \begin{pmatrix} D_F'' & \alpha^{0,1} \\ 0 & D_{F^\perp}'' \end{pmatrix}$$

and

$$\phi = \begin{pmatrix} \phi_F & \beta^{1,0} \\ 0 & \phi_{F^\perp} \end{pmatrix}.$$

In this way we are able to compare the curvatures of $E$ and $F$ for the given metric $h$:

$$(\nabla_h^F)^2 = \pi_F (\nabla_h^E)^2 \pi_F + \alpha \wedge \alpha^* - \beta \wedge \beta^*.$$

Applying the operator $\Lambda$ we obtain

$$\int_X c_1(F) \wedge \omega^{k-1} = \int_X \Lambda (\nabla_h^F)^2 = \int_X (\Lambda \pi_F (\nabla_h^E)^2 \pi_F - |\alpha|^2 - |\beta|^2)$$

$$= \text{rk}\,(F) \mu(E) - \int_X (|\alpha|^2 + |\beta|^2).$$

from which our claim follows. □

The converse of this result is the following theorem of Hitchin [**72**] and Simpson [**115**]. Earlier versions, without the Higgs field, were proved by Narasimhan–Seshadri (in the case of bundles over curves), Donaldson, and Uhlenbeck–Yau.

THEOREM 7.20. *If $(E, \phi)$ is a stable Higgs bundle, then it admits a solution to the HHYM equation. This solution is unique up to scalar multiplication.*

The full proof is too long to describe here, but we will at least try to indicate a few of the main ideas of the proof given by Hitchin in the particular case where $X$ is a curve, and $E$ is a rank two vector bundle with $c_1(E) = 0$.

Fix a smooth bundle $E$ and a metric $h$ on it. Consider pairs $(D_E'', \phi)$.
*Problem:* Find a solution to the HHYM equation in a given $\mathcal{G}^\mathbb{C}$–orbit of Higgs bundle structures, where $\mathcal{G}^\mathbb{C}$ is the group of complex gauge automorphisms of $E$.

For this, select a reference Higgs structure $(D_E'', \phi)$, and choose a sequence of complex gauge transformations $g_i$ which gives a minimising sequence for the curvature norm $\|F^{\nabla_{g_i}}\|_{L^2}^2$, where $\nabla_g$ is the connection associated with the fixed metric and the Higgs bundle obtained by transforming the original Higgs bundle structure by $g$. The proof has two steps:

*Step 1.* Show that there is a weak limit of $\nabla_{g_i}$ in the Sobolev space of smooth connections completed in the $L^{2,1}$–norm, possibly after modifying the $g_i$ by unitary gauge transformations.

*Step 2.* Show that this limit is a smooth connection in the original gauge orbit.

The first step is based in an essential way on a theorem of Uhlenbeck, which asserts that, in the case of a Riemann surface, an $L^2$–bound on the curvature of a sequence of Hermitian connections implies the existence of a sequence of unitary gauge transformations such that the modified sequence of connections has a subsequence which converges weakly in the Sobolev $L^{2,1}$ sense. The main point is to show that the bound on the curvatures of the nonunitary connections $\nabla_{g_i}$ implies a bound on the curvature of the unitary part of the connection.

In the second step, one exploits the semicontinuity property of dimensions of spaces of holomorphic endomorphisms to conclude that there is a nontrivial Higgs bundle morphism from the limiting Higgs bundle to the initial one (since there is one from each of the Higgs bundles in the sequence). This morphism is either an isomorphism, in which case we are done, or its image is a nontrivial proper Higgs subsheaf of the initial Higgs bundle. At that point, one can exploit the assumption of stability to exclude the second possibility.

**2.2. Examples and applications.** Let us describe some examples, due to Hitchin [**72**], of Higgs bundles over a compact Riemann surface $X$. We denote the canonical bundle of $X$ by $K$.

EXAMPLE 7.21. Let $E$ be the rank three bundle $E = K \oplus \mathcal{O} \oplus K^*$, and consider the Higgs field

$$\phi = \begin{pmatrix} 0 & 0 & 0 \\ 1 & 0 & 0 \\ 0 & 1 & 0 \end{pmatrix},$$

where each nonzero entry represents the identity map

$$\mathrm{Id} \colon K \longrightarrow K \otimes \mathcal{O} \cong K$$
$$\mathrm{Id} \colon \mathcal{O} \longrightarrow K \otimes K^* \cong \mathcal{O}.$$

The bundle $E$ has degree zero, and its only proper Higgs subbundle is $K^*$. Therefore $(E, \phi)$ is stable if and only if $K^*$ has strictly negative degree, equivalently if and only if $X$ has genus $g \geq 2$. We will assume this for the rest of this section.

By Theorem 7.20 $E$ admits an HHYM metric. Thus it arises from a flat bundle, and we get from it an associated harmonic map

$$\tilde{h} \colon \tilde{X} \longrightarrow D$$

from the universal cover of $X$ to the Poincaré disk. This map is a conformal diffeomorphism, because the Higgs field is the $(1,0)$–component of its differential, and $\phi$ takes values in the part of the endomorphism bundle corresponding to the $(1,0)$ piece of the tangent space of the disk and vanishes nowhere. Therefore, we recover the uniformisation theorem for hyperbolic compact Riemann surfaces.

EXAMPLE 7.22. Let $X$ and $E$ be as in the previous example, and consider a Higgs field

$$\phi = \begin{pmatrix} 0 & \alpha & 0 \\ 1 & 0 & \alpha \\ 0 & 1 & 0 \end{pmatrix}$$

determined by

$$Id\colon K \longrightarrow K \otimes \mathcal{O} \cong K$$
$$Id\colon \mathcal{O} \longrightarrow K \otimes K^* \cong \mathcal{O}$$
$$\alpha\colon \mathcal{O} \longrightarrow K \otimes K$$
$$\alpha\colon K^* \longrightarrow K \otimes \mathcal{O}$$

Such Higgs fields are parametrised by the vector space of quadratic differentials $\alpha \in H^0(X, K^{\otimes 2})$, which is diffeomorphic to the Teichmüller space of the corresponding genus, and also to the space of flat $PSL(2, \mathbb{R})$–bundles with the same Euler characteristic as the uniformising bundle.

EXAMPLE 7.23. In general, one can consider rank three bundles

$$E = L \oplus \mathcal{O} \oplus L^*$$

for a given line bundle $L$, and Higgs fields

$$\phi = \begin{pmatrix} 0 & \beta & 0 \\ \alpha & 0 & \beta \\ 0 & \alpha & 0 \end{pmatrix}$$

determined by sections $\alpha \in H^0(X, K \otimes L^*)$ and $\beta \in H^0(X, K \otimes L)$.

Provided that $\alpha \neq 0$, $E$ is stable if and only if $\deg L > 0$. A necessary condition to get a flat bundle is that $\deg L < \deg K$. The Higgs bundle $(E, \phi)$ is determined up to equivalence by $\beta$ and the divisor of $\alpha$, so the moduli space of flat bundles corresponding to $\deg(L) = d$ is homeomorphic to a vector bundle of rank $(g - 1 + d)$ over $Sym^{2g-2-d}X$.

## 3. Hyperkähler structures and complete integrability

We will discuss the case of compact Riemann surfaces, although many of the results discussed here are also valid for compact Kähler manifolds of arbitrary dimension. Let $X$ be a compact Riemann surface with metric, and let $E \to X$ be a smooth $\mathbb{C}$-vector bundle, endowed with a Hermitian metric $h$.

We will denote by $\mathcal{A}^{\mathbb{C}}$ the space of all linear connections on this vector bundle. The Hermitian metric $h$ together with the Kähler metric on $X$ induces an inner product on 1-forms with values in $End(E)$, and thus a flat Riemannian metric on the affine space $\mathcal{A}^{\mathbb{C}}$. Let $\mathcal{G}$ be the group of unitary gauge automorphisms of $E$. This acts by isometries on $\mathcal{A}^{\mathbb{C}}$ by

$$\nabla \longmapsto g \nabla g^{-1} = \nabla + (\nabla g) g^{-1} \ .$$

A remarkable property of this action is that it admits three different invariant symplectic structures. These symplectic structures arise from complex structures on $\mathcal{A}^{\mathbb{C}}$ coming from two sources.

DEFINITION 7.24. (i) The complex structure $I$ is given by $I = 1 \otimes i$ acting on $\Gamma(T^*X \underset{\mathbb{R}}{\otimes} End(E))$. It comes from the complex structure of $E$.
(ii) The complex structure $J$ is given by $J = j \otimes \tau$, where $j\colon TX \to TX$ is the complex structure of $X$ and $\tau\colon End(E) \to End(E)$ is the adjoint map.
(iii) The complex structure $K$ is defined by $K = IJ = -JI$.

These three complex structures are parallel and orthogonal with respect to the metric $h$, hence they induce Kähler structures on $\mathcal{A}^\mathbb{C}$, all preserved by the unitary gauge group $\mathcal{G}$.

DEFINITION 7.25. Riemannian manifolds with three parallel orthogonal complex structures satisfying the quaternion relations (like the above $I$, $J$ and $K$ on $\mathcal{A}^\mathbb{C}$) are called *hyperkähler manifolds*.

**3.1. Moment maps.** Let $(M, \omega)$ be a symplectic manifold, and $G \times M \to M$ a symplectic group action. Denote by $\mathfrak{g}$ the Lie algebra of $G$ and by $Vect(M, \omega)$ the space of those smooth vector fields on $M$, which infinitesimally preserve $\omega$. The group action induces a homomorphism
$$\mathfrak{g} \longrightarrow Vect(M, \omega) .$$
Moreover, the symplectic form $\omega$ gives an identification of vector fields and 1-forms. The vector fields which preserve the symplectic form are identified with closed 1–forms; those vector fields which are identified with exact 1–forms are called Hamiltonian, and will be denoted by $Ham(M, \omega)$. Thus there is an exact sequence
$$(11) \qquad 0 \longrightarrow \mathbb{R} \longrightarrow \mathcal{C}^\infty(M) \longrightarrow Ham(M, \omega) \longrightarrow 0$$
where the last map is the composition
$$f \xrightarrow{d} df \xrightarrow{\omega(\cdot, \cdot)} \xi_f .$$
The function space $\mathcal{C}^\infty(M)$ admits a Lie algebra structure defined by the *Poisson bracket* given by
$$\{f, g\} = \omega(\xi_f, \xi_g) .$$
The Poisson bracket makes the exact sequence (11) an exact sequence of Lie algebras. The group action is said to be *Hamiltonian* if the homomorphism from $\mathfrak{g}$ to $Vect(M, \omega)$ factors through $Ham(M)$. If this map admits a lifting to $\mathcal{C}^\infty(M)$, it induces a map
$$\Psi \colon M \longrightarrow \mathfrak{g}^* .$$
This map is called a *moment map* for the action. Given such a map, the quotient $\Psi^{-1}(0)/G$ is the *Marsden–Weinstein* quotient of $M$ by $G$. Under appropriate hypotheses (freedom of the action, zero a regular value of the moment map), the Marsden–Weinstein quotient $\Psi^{-1}(0)/G$ is a symplectic manifold. If $(M, \omega)$ is the underlying symplectic manifold of a Kähler manifold and $G$ preserves the complex structure of $M$, then $\Psi^{-1}(0)/G$ is also Kähler.

The Marsden–Weinstein quotient has dimension $\dim \Psi^{-1}(0)/G = \dim M - 2\dim G = \dim M - \dim G_\mathbb{C}$, which suggests some parallel with the quotient by the complexified group $G_\mathbb{C}$. Indeed, if $M$ is projective algebraic and $G$ is compact linear, then $\Psi^{-1}(0)/G$ is equivalent to the Geometric Invariant Theory quotient of $M$ by $G_\mathbb{C}$. Why this is so may be seen roughly as follows.

Choose a line bundle $L$ with a connection $\nabla$ such that $\nabla^2 = 2\pi i \omega$. The existence of a moment map is equivalent to the existence of a lifting of the action of $G$ from $M$ to the pair $(L, \nabla)$. If $L$ arises from a projective embedding of $M$, this is a linearisation of the action.

EXAMPLE 7.26. If $G$ acts linearly on $V$, preserving a Hermitian form $h$, a moment map is given by
$$\Psi(v)(A) = h(Av, v) .$$

This is the differential of the norm function $\|v\|^2$ along the $G$–orbit. The moment map has a zero along a given orbit if and only if the norm achieves a minimum along the orbit.

THEOREM 7.27 (Kempf–Ness). *If an orbit is stable in the previous example, then $\Psi$ has a zero on it.*

In fact, if the action is essentially free, then existence of zeroes and stability are equivalent.

If $M$ is hyperkähler, it has three symplectic structures whose corresponding forms are $\omega_I$, $\omega_J$ and $\omega_K$. When moment maps exist for the three structures, one can consider the combined map:
$$\mu = \Psi_I \otimes I + \Psi_J \otimes J + \Psi_K \otimes K \colon M \longrightarrow \mathfrak{g}^* \otimes \operatorname{Im} \mathbb{H}.$$

DEFINITION 7.28. The *hyperkähler quotient* of $M$ is
$$\mu^{-1}(0)/G.$$

THEOREM 7.29 (Hitchin–Karlhede–Lindstöm–Rocek). *The quotient $\mu^{-1}(0)/G$ is a hyperkähler manifold.*

This Theorem can be applied to $\mathcal{A}^{\mathbb{C}}$. For the action of $\mathcal{G}$ on $\mathcal{A}^{\mathbb{C}}$, the moment maps are, up to normalisation,
$$\Psi_I = D^*\theta$$
$$\Psi_J = \operatorname{Im} F^\nabla$$
$$\Psi_K = \operatorname{Re} F^\nabla.$$

Thus $\mu^{-1}(0)$ is the subspace of harmonic flat bundles on $\mathcal{A}^{\mathbb{C}}$, and Theorem 7.29 implies in this case:

THEOREM 7.30 (Hitchin [72], Fujiki [49]). *The smooth part of the moduli space $\mathcal{M}$ is a hyperkähler manifold.*

We will not go into the proof of these theorems, but instead give some reasons to expect a holomorphic symplectic structure on $\mathcal{M}$:
(i) *Poincaré duality*: Consider the moduli space of flat bundles $\mathcal{M}_B$. Its tangent space at a point represented by $E$ is
$$T\mathcal{M}_B = H^1(X, End(E)).$$
Poincaré duality gives a nondegenerate skew–symmetric pairing
$$H^1(X, End(E)) \times H^1(X, End(E)) \longrightarrow \mathbb{C}$$
$$(\alpha, \beta) \longmapsto \int_X \operatorname{Tr}(\alpha \wedge \beta)$$
(In the case of higher dimensional manifolds this pairing can be extended by adding a $\wedge \omega^{n-2}$ factor.) Goldman [50] proved that this pairing induces a closed 2–form by using moment maps.
(ii) *The moduli space of Higgs bundles*: The moduli space of Higgs bundles is denoted by $\mathcal{M}_{Dol}$. Let $\mathcal{S}$ be the moduli space of stable holomorphic bundles over $X$. The moduli space $\mathcal{M}_{Dol}$ contains a vector bundle over $\mathcal{S}$, whose fiber over $[E] \in \mathcal{S}$ is
$$H^0(X, \Omega^1 \otimes End(E)) \cong \left(H^1_{Dol}(X, End(E))\right)^* = (T_{[E]}\mathcal{S})^*$$

Therefore, the cotangent bundle $T^*\mathcal{S}$ is contained in the Higgs bundle moduli space $\mathcal{M}_{Dol}$.

Define now a norm map
$$\varepsilon : \mathcal{A}^{\mathbb{C}} \longrightarrow \mathbb{R}$$
$$\nabla \longmapsto \|\theta\|^2 \ .$$

The norm $\varepsilon$ is invariant under the action of the unitary gauge group $\mathcal{G}$, so it descends to the hyperkähler quotient $\mathcal{M}$. It turns out that the Hamiltonian vector field $\xi$ associated to $\varepsilon$ by $\omega_J$ is the $S^1$-action on $\mathcal{M}$. The gradient flow of $\varepsilon$ is the $\mathbb{R}^*$ part of the $\mathbb{C}^*$-action on $\mathcal{M}$ constructed in the previous section.

QUESTION 7.31. *Does this gradient flow converge as the time tends to infinity? Equivalently, does the limit of $(E, \lambda\theta)$ as $\lambda$ tends to zero exist in $\mathcal{M}$?*

In the case when $X$ is a smooth projective algebraic variety, the answer to this question turns out to be affirmative.

THEOREM 7.32 (Hitchin, Simpson). *Let $(E, \phi)$ be a Higgs bundle over a smooth projective algebraic manifold $X$. Then*
$$\lim_{\lambda \to 0}(E, \lambda\phi)$$
*exists in $\mathcal{M}$.*

This follows from a stronger result. To formulate it, we need to define the Hitchin map on the moduli space of Higgs bundles. This is a map $H$ from the moduli space $\mathcal{M}_{Dol}$ of Higgs bundles of rank $k$ to the space of holomorphic differentials on $X$ by:
$$H(E, \theta) = Tr\,\theta \oplus Tr\,\theta^2 \oplus \ldots \oplus Tr\,\theta^k \ .$$

THEOREM 7.33 (Hitchin [**72**], Simpson [**118**] [**119**]). *The map*
$$H \colon \mathcal{M} \longrightarrow \oplus_{i=1}^{k} H^0(X, Sym^i(\Omega^1))$$
*is proper.*

There are two ways of approaching the proofs of these theorems. One way, used by Hitchin in the case of curves, is gauge theoretic in nature, and is essentially a generalisation of his proof of the existence of HHYM connections. The second approach, due to Simpson, is algebro–geometric, and is a generalisation of ideas used to prove the properness of moduli spaces of stable bundles.

There is an interesting interplay between the map $H$ and the symplectic structure on $\mathcal{M}$.

THEOREM 7.34 (Hitchin [**73**]). *For Riemann surfaces, the functions induced on $\mathcal{M}_{Dol}$ by pullback along $H$ Poisson commute with each other. The generic fiber of $H$ is an Abelian variety and the Dolbeault moduli space $\mathcal{M}_{Dol}$ is algebraically completely integrable with respect to the functions induced by $H$.*

## 4. Applications

In this Section, we take up two applications of the ideas of non–Abelian Hodge theory. The first application is an analogue of the formality result discussed in Chapter 3, and the second is Reznikov's proof of Bloch's conjecture on the secondary characteristic classes of flat bundles over projective varieties.

### 4.1. Formality of the de Rham complex.

Let $(E, \nabla, h)$ be a harmonic flat bundle. We have discussed so far two complexes associated to it: the de Rham complex

$$0 \longrightarrow \mathcal{E}^0(X, E) \xrightarrow{\nabla} \mathcal{E}^1(X, E) \xrightarrow{\nabla} \ldots$$

and its Dolbeault analogue

$$0 \longrightarrow \mathcal{E}^0(X, E) \xrightarrow{\mathcal{D}''} \mathcal{E}^1(X, E) \xrightarrow{\mathcal{D}''} \ldots$$

The question we seek to answer is how these two complexes relate to each other. In the case when $E$ is a trivial line bundle, the above Dolbeault complex is the classical one, and it is well-known that it computes the same cohomology as the de Rham complex. In general, the *Kähler identities* still hold.

PROPOSITION 7.35 (Deligne, Simpson [116]). *For any harmonic bundle $(E, \nabla, h)$ the following identities hold:*

$$(\mathcal{D}')^* = i[\Lambda, \mathcal{D}'']$$
$$(\mathcal{D}'')^* = -i[\Lambda, \mathcal{D}'] \ .$$

As a consequence of these Kähler identities there are some identities between Laplacians:

COROLLARY 7.36. *The following identities hold:*

$$\square^\nabla = 2\square^{\mathcal{D}''} = 2\square^{\mathcal{D}'} \ .$$

Hence we deduce an isomorphism of cohomology groups:

$$H^k(X, E) \cong H^k_{\mathcal{D}''}(X, E) \ .$$

The complex relevant to the deformation theory of a flat bundle is the de Rham complex for the endomorphism bundle:

$$(\mathcal{E}^*(X, End(E)), \nabla)) \ .$$

An $End(E)$–valued 1–form $\eta$ on $X$ is, to first order, a direction in which the flat connection can be deformed if $\eta$ is closed; higher order obstructions to extending it to an actual deformation occur in the second cohomology group of this complex. The direct sum of the terms of this complex is a differential graded Lie algebra (DGLA), with product given by the bracket of forms with values in $End(E)$, and differential the exterior derivative.

To any DGLA $(\mathcal{E}^*, d)$ we can associate a deformation theory, by which we mean that we consider DGLAs with a perturbed boundary $(\mathcal{E}^*, d + A)$, where $A$ is an element of $\mathcal{E}^1$ such that $(d + A)^2 = 0$. There is a parametrising variety for the solutions $A$, and the subspace of elements of $\mathcal{E}^0$ which are invertible and preserve $d$ acts on this solution variety. This is a groupoid. We can consider the formal completion of this groupoid around $d$. The deformation theories of two DGLAs are said to agree if there is an equivalence between the corresponding formally completed groupoids. (The reader is advised to consult the paper of Goldman and Millson [51] for a more detailed discussion of this.)

There is a twisted analogue of the results discussed in Chapter 3 on the formality of the algebra of differential forms on a compact Kähler manifold. Let us recall the notion of quasi–isomorphism (which was called weak equivalence in Chapter 3):

DEFINITION 7.37. The DGLAs $(\mathcal{E}_1^*, d_1)$ and $(\mathcal{E}_2^*, d_2)$ are *quasi–isomorphic* if there exists a finite chain

$$(\mathcal{E}_1^*, d_1) \to (\mathcal{A}_1, \delta_1) \leftarrow (\mathcal{A}_2, \delta_2) \to \cdots \leftarrow (\mathcal{E}_2^*, d_2)$$

where every map is a DGLA map inducing an isomorphism of the cohomology rings.

The following (proved by Simpson [116]) is the twisted analogue of Theorem 3.13 of Deligne–Griffiths–Morgan–Sullivan.
*Claim:* The DGLA $(\mathcal{E}^*(X, End(E)), \nabla))$ is formal, i.e., it is quasi–isomorphic to its cohomology algebra.

FACT 7.38 (Deligne, Schlessinger, Stasheff, Goldman–Millson [51]). Quasi-isomorphic DGLAs give the "same" deformation theory, in the sense described above.

Let us now show the formality of $(\mathcal{E}^*(End(E)), \nabla)$. That is, we must find a chain as in Definition 7.37 linking the complex and its cohomology. Consider the intermediate complex

$$\begin{array}{ccc} (\ker(\mathcal{D}'), \mathcal{D}'') & \stackrel{\nu}{\hookrightarrow} & (\mathcal{E}^*(End(E)), \nabla) \\ \downarrow & & \\ (H^*(X, End(E)), 0) & & \end{array}$$

The vertical arrow is the map determined by assigning to a form which is annihilated by $\mathcal{D}'$ its class in the $\mathcal{D}'$–cohomology, which, by the discussion above, can be identified with the de Rham cohomology of the flat bundle $End(E)$. We must check that both maps in the diagram induce isomorphisms in cohomology. It actually suffices to show that every class in the cohomology of the intermediate complex $(\ker(\mathcal{D}'), \mathcal{D}'')$ is represented by a harmonic form.

Take a cocycle $\alpha \in \mathcal{E}^k(X, End(E))$ such that

$$\mathcal{D}'\alpha = \mathcal{D}''\alpha = 0 \ .$$

Suppose that its cohomology class $[\alpha] \in H_{\mathcal{D}''}^k(X, End(E))$ vanishes, i.e., $\alpha = \mathcal{D}''\beta$. The key point in our argument is a version of a classical result:

LEMMA 7.39. *Principle of two types (without types!)*

$$\ker(\mathcal{D}') \cap \ker(\mathcal{D}'') \cap (\mathrm{Im}\,(\mathcal{D}') + \mathrm{Im}\,(\mathcal{D}'')) = \mathrm{Im}\,(\mathcal{D}'\mathcal{D}'')$$

Thus if $\alpha = \mathcal{D}''\beta$, then as $\mathcal{D}'\alpha = \mathcal{D}''\alpha = 0$, by the principle of two types $\alpha = \mathcal{D}''(\mathcal{D}'\gamma)$, so it is a coboundary in the complex $(\ker(\mathcal{D}'), \mathcal{D}'')$, and the map $\nu^*$ (i.e., the map induced by $\nu$ on cohomology) is injective. It is clearly surjective, so it is an isomorphism. The same holds for the map into $(H^*(X, End(E)), 0)$, establishing formality.

As a consequence, the deformation theories given by $(H^*(X, End(E)), 0)$ and $(\mathcal{E}^*(End(E)), \nabla)$ are the same.

COROLLARY 7.40 (Goldman–Millson [51], Simpson [116]). *For a harmonic flat bundle $E$, the only obstruction to extending a class $\alpha \in H^1(X, End(E))$ to a deformation of flat bundles is the condition $[\alpha, \alpha] = 0$ in $H^2(X, End(E))$.*

This shows that the representation varieties of Kähler groups have at worst quadratic singularities at reductive representations. More precisely, for any reductive representation, there is a neighbourhood in the representation variety which is analytically equivalent to a cone in an affine space defined by homogeneous quadratic equations. For fundamental groups of reducible algebraic varieties, this is not true, because all finitely presentable groups occur, cf. subsection 2.3 in Chapter 1.

**4.2. Bloch's conjecture on Cheeger–Chern–Simons classes of flat bundles.** Let $X$ be a smooth projective algebraic variety over $\mathbb{C}$, and $(E, \nabla)$ a flat vector bundle over $X$. The underlying algebraic vector bundle has Chern classes in the Chow ring of algebraic cycles. Bloch [12] formulated a conjecture which says that those Chern classes should lie in the "lowest stratum" of the Chow group. By applying the Abel–Jacobi map, he obtained a conjecture about the Cheeger–Chern–Simons classes of flat bundles over algebraic varieties, which has recently been proved by Reznikov.

Consider the frame bundle $P$ of $E$, with connection form $\theta$. If $p$ is an invariant polynomial of degree $k$ on $\mathfrak{gl}(n, \mathbb{C})$ then $p(F^\nabla)$ is a form of degree $2k$ on $P$. The form $p(F^\nabla)$ descends to $X$, and represents a characteristic class of $E$.

PROPOSITION 7.41 (Chern–Simons [32]). *There is a canonical $(2k-1)$-form $\tilde{p}(\nabla)$ on $P$ satisfying*
$$d\tilde{p}(\nabla) = p(F^\nabla) .$$
*If the connection $\nabla$ is flat, then $\tilde{p}(\nabla)$ is closed.*

REMARK 7.42. The formula for $\tilde{p}$ is
$$\tilde{p}(\nabla) = h \int_0^1 p(\theta, \varphi_t, \ldots, \varphi_t) dt ,$$
where $\varphi_t = tF^\nabla + \frac{1}{2}(t^2 - t)[\theta, \theta]$.

The form $\tilde{p}(\nabla)$ does not generally arise as the pullback of a form on $X$ (unlike $p(\nabla)$), but it can be shown [31] that it gives rise to a canonical class in $H^{2k-1}(X, \mathbb{C}/\mathbb{Z})$, which will also be denoted by $\tilde{p}(\nabla)$.

EXAMPLE 7.43. For a flat circle bundle, the form $\tilde{c}_1(\nabla)$ is the connection form, while the corresponding cohomology class in $H^1(X, \mathbb{C}/\mathbb{Z}) = H^1(X, \mathbb{C}^*)$ is that determined by the usual identification of the latter group with isomorphism classes of flat $\mathbb{C}^*$ bundles.

We can now state Bloch's conjecture:

CONJECTURE 7.44. *If $X$ is projective, the classes $\tilde{c}_k(\nabla)$ are torsion for $k > 1$.*

The rest of this Chapter will be devoted to describing Reznikov's proof [106] of this conjecture.

For $k > 1$, the classes $\tilde{c}_k(\nabla)$ are rigid under deformation. In the component of the moduli space of flat bundles containing $[\nabla]$ there is some connection $\nabla_1$ whose monodromy admits coefficients in an algebraic number field $K$, in fact in the ring of integers $\mathcal{O}_S$ of $K$ with finitely many primes inverted.

Thus $\tilde{c}_k(\nabla_1) \in H^{2k-1}(X, \mathbb{C}/\mathbb{Z})$ is the pullback of a class $\psi$ in $H^{2k-1}(SL(n, \mathcal{O}_S), \mathbb{C}/\mathbb{Z})$, which is stable as the rank $n$ tends to infinity. Let $\Gamma_n = SL(n, \mathcal{O}_S)$. We get in this manner a map of exact sequences

$$\begin{array}{ccccccc}
H^{2k-1}(X, \mathbb{Z}) & \longrightarrow & H^{2k-1}(X, \mathbb{C}) & \longrightarrow & H^{2k-1}(X, \mathbb{C}/\mathbb{Z}) & \longrightarrow & H^{2k}(X, \mathbb{Z}) \\
\uparrow & & \uparrow & & \uparrow & & \uparrow \\
H^{2k-1}(\Gamma_n, \mathbb{Z}) & \longrightarrow & H^{2k-1}(\Gamma_n, \mathbb{C}) & \longrightarrow & H^{2k-1}(\Gamma_n, \mathbb{C}/\mathbb{Z}) & \longrightarrow & H^{2k}(\Gamma_n, \mathbb{Z})
\end{array}$$

## 4. APPLICATIONS

To prove the Conjecture, it suffices to show that the image
$$H^{2k-1}(\Gamma_n, \mathbb{C}/\mathbb{Z}) \longrightarrow H^{2k-1}(X, \mathbb{C}/\mathbb{Z})$$
is torsion. For this, it is enough to show that
$$H^{2k-1}(\Gamma_n, \mathbb{C}) \longrightarrow H^{2k-1}(X, \mathbb{C})$$
has zero image. For any given $k$, the classes with which we are concerned are stable as $n$ gets large, so we need only consider the stable cohomology in the above diagram of exact sequences. Thus we must check that the image
$$H^{2k-1}(\Gamma_n, \mathbb{C}) \longrightarrow H^{2k-1}(X, \mathbb{C})$$
is zero for $n$ sufficiently large. Assume that $K$ has no real embeddings, and let $\sigma : K \to \mathbb{C}$ be an embedding. We can obtain classes in $H^{2k-1}(\Gamma_n, \mathbb{C})$ by pulling back invariant forms on the symmetric space $SL(n, \mathbb{C})/SU(n)$. The invariant forms on $SL(n, \mathbb{C})/SU(n)$ are found by taking the traces $\operatorname{Tr} \theta^{2k-1}$. The proof is concluded now by applying either of the two following arguments plus a theorem of Borel.

*First argument*: Consider the image of $\operatorname{Tr} \theta^{2k-1}$ on $X$: the Siu–Sampson formula says that
$$\theta^{1,0} \wedge \theta^{1,0} = 0 \ .$$
For $k > 1$, every summand in $\operatorname{Tr}(\theta^{1,0} + \theta^{0,1})^{2k-1}$ has at least two factors $\theta^{1,0}$ or $\theta^{0,1}$, so the form vanishes.

*Second argument*: Deform $(E, \nabla)$ to a VHS, and then observe that the only nontrivial invariant forms on a period domain occur in even degree, while $\operatorname{Tr} \theta^{2k-1}$ is of odd degree.

After either argument, the proof is finished by the following result:

THEOREM 7.45 (Borel [15], [17]). *If $K, \mathcal{O}_S$ are as above, then, for $n$ sufficiently large compared to $k$, the groups $H^{2k-1}(SL(n, \mathcal{O}_S), \mathbb{C})$ are generated by pullbacks of invariant forms on the symmetric spaces $SL(n, \mathbb{C})/SU(n)$, ranging over all embeddings $\sigma : K \hookrightarrow \mathbb{C}$.*

CHAPTER 8

# Positive results for infinite groups

## 1. Introduction

The examples of infinite Kähler groups that one can write down easily are very restricted. The aim of this Chapter is to explain an idea that allows one to generate a number of new nontrivial examples. The idea is to find pairs $V \subset \bar{V} \subset \mathbb{P}^N$ with $\bar{V}$ and $\bar{V} \setminus V$ projective varieties, such that there exists a projective subvariety $M \subset V$ whose inclusion map $\nu : M \hookrightarrow V$ induces an isomorphism of fundamental groups $\pi_1(M) \cong \pi_1(V)$.

One of the important applications of this idea which we shall discuss is the construction of Kähler groups which are not residually finite. A group $\Gamma$ is said to be *residually finite* if every element of $\Gamma \setminus \{1\}$ has a non–trivial image in a finite quotient of $\Gamma$, equivalently,

$$\bigcap_{\substack{\Delta \subset \Gamma \\ finite\ index}} \Delta = \{1\} \ .$$

Strictly speaking, we should take the intersection over all normal subgroups $\Delta$ of finite index in $\Gamma$. But it is easy to see that every subgroup of finite index contains a normal subgroup of finite index, thus the two definition agree. See Appendix A for further discussion.

Issues of residual finiteness first appeared in algebraic geometry in Zariski's 1934 book on algebraic surfaces [**137**]. There the fundamental group of the complement of a plane curve, $\Gamma = \pi_1(\mathbb{C}P^2 \setminus C)$, is studied. Zariski remarks that what one can calculate with algebraic coverings is not $\Gamma$ itself, but the quotient

$$\Gamma / \Big( \bigcap_{\substack{\Delta \subset \Gamma \\ f.i.}} \Delta \Big) \ .$$

This motivated the following question:

QUESTION 8.1 (Zariski, 1930s). *Is there a (finitely presentable) group $\Gamma$ such that*

$$\bigcap_{\substack{\Delta \subset \Gamma \\ f.i.}} \Delta \neq \{1\} \quad ?$$

The restriction to finitely presentable groups is natural here, since the question concerns fundamental groups of spaces homotopy equivalent to finite complexes. For infinitely generated groups one has of course the example of the alternating group on infinitely many letters, which is a simple group.

As free products cannot be Kähler groups, cf. Chapter 4, it is interesting to note that Zariski gave an example of a curve in $\mathbb{C}P^2$ whose complement has fundamental group $\mathbb{Z}_2 * \mathbb{Z}_3$. The curve is defined in homogeneous coordinates by $p_2^3 = p_3^2$, where $p_i$ is a generic homogeneous polynomial of degree $i$.

The first example of a finitely generated group which is not residually finite was given by B. Neumann [**99**] in 1950. The first examples of finitely presentable, non–residually finite groups were given by G. Higman [**69, 70**] in 1951. The second paper introduces the Higman 4–group discussed in Chapters 1 and 6. Recall that this group is infinite, and it contains no proper subgroup of finite index. It is not a Kähler group, by the results of Gromov–Schoen [**62**] discussed in Chapter 6.

Neumann's example, as well as the example in the first paper of Higman, are of non–Hopfian groups, i.e., groups isomorphic to a proper quotient of themselves

$$\Gamma \xrightarrow{\cong} \Gamma/\Delta \ .$$

These papers do not mention residual finiteness, but it was known at the time (as a consequence of results of Malcev) that residually finite groups are Hopfian. The second paper of Higman is directly concerned with the issue of residual finiteness.

We should also mention that residually finite groups have solvable word problem, cf. IV 4.6 of [**88**]. Thus any group with unsolvable word problem provides an example of a non–residually finite group.

Given these developments in group theory, the reasonable question to ask is whether fundamental groups of algebraic varieties, or specific classes of algebraic varieties, are residually finite. One could consider, for instance complements of plane curves as considered by Zariski, or smooth varieties, or projective varieties, etc. Such questions were posed explicitly by Abhyankar in Remark 15 of [**1**], and by Serre (folklore, and the Open Problem 1.17 mentioned in Chapter 1).

At present one does not know if there are complements of curves in the projective plane with non–residually finite fundamental group. But in recent years examples of projective (and quasi–projective) varieties with non–residually finite fundamental group have appeared. As we shall explain in this Chapter, all the examples depend on properties of lattices in non–linear Lie groups.

Recall that Malcev proved that finitely generated linear groups are residually finite. But his arguments do not apply to finitely generated subgroups of non–linear Lie groups. In fact, examples of non–residually finite lattices in non–linear Lie groups have been given by Millson, Serre, Deligne, Raghunathan, Prasad, and presumably others.

The following is perhaps the simplest example:

EXAMPLE 8.2.* (Millson [**93**]) Let $\widetilde{SL}(3,\mathbb{R})$ be the universal cover of $SL(3,\mathbb{R})$, and let $\Gamma$ be the pre-image of $SL(3,\mathbb{Z})$ in $\widetilde{SL}(3,\mathbb{R})$:

$$\begin{array}{ccc} \Gamma & \hookrightarrow & \widetilde{SL}(3,\mathbb{R}) \\ \downarrow & & \downarrow \\ SL(3,\mathbb{Z}) & \hookrightarrow & SL(3,\mathbb{R}) \end{array}$$

The group $\Gamma$ is not residually finite. More precisely, it is an extension

$$1 \longrightarrow \bigcap_{\substack{H \subset \Gamma \\ f.i.}} H \cong \mathbb{Z}/2\mathbb{Z} \longrightarrow \Gamma \longrightarrow SL(3,\mathbb{Z}) \longrightarrow 1 \ .$$

The following example is perhaps the simplest one which is relevant to Kähler groups:

---

*Added in proof: In $SL(3,\mathbb{Z})$ replace $\mathbb{Z}$ by the integers in $\mathbb{Q}(\sqrt{7})$.

EXAMPLE 8.3. (Deligne [**39**]) Let $\widetilde{Sp}(2n,\mathbb{R})$ denote the universal cover of the symplectic group $Sp(2n,\mathbb{R})$ (which preserves a symplectic form on $\mathbb{R}^{2n}$), and suppose that $n > 1$. Define $\Gamma$ to be the pre–image of $Sp(2n,\mathbb{Z})$ in $\widetilde{Sp}(2n,\mathbb{R})$:

$$\begin{array}{ccc} \Gamma & \hookrightarrow & \widetilde{Sp}(2n,\mathbb{R}) \\ \downarrow & & \downarrow \\ Sp(2n,\mathbb{Z}) & \hookrightarrow & Sp(2n,\mathbb{R}) \end{array}$$

Then the group $\Gamma$ is not residually finite. More precisely, it is a central extension

$$1 \longrightarrow \mathbb{Z} \longrightarrow \Gamma \longrightarrow Sp(2n,\mathbb{Z}) \longrightarrow 1 ,$$

and

$$\bigcap_{\substack{H \subset \Gamma \\ f.i.}} H = 2\mathbb{Z} .$$

## 2. The first construction

We first explain, in the context of linear groups, a geometric idea that will also be used later to construct non–residually finite Kähler groups. Consider the linear group $SU(1,n,\mathbb{Z}[i])$, defined as

$$SU(1,n,\mathbb{Z}[i]) = \{A \in SL(n+1,\mathbb{Z}[i]) \mid A^*h = h\}$$

where $h$ is the bilinear form $h(z) = |z_0|^2 - |z_1|^2 - \cdots - |z_n|^2$. This group is a non–cocompact lattice in $SU(1,n)$.

Let $\Gamma \subset SU(1,n,\mathbb{Z}[i])$ be a torsion–free lattice of finite index, and denote by $B^n$ the unit ball in $\mathbb{C}^n$. The quotient

$$V = \Gamma \backslash B^n$$

is a smooth open variety of finite volume. There is a natural projective compactification $V \subset \bar{V} = V \cup \{P_1, \ldots, P_k\} \subset \mathbb{P}^N$ which is the union of $V$ with a finite number of cusps. The level sets obtained by inter ecting $V$ with small horospheres centreed on the cusps $P_i$ are nilmanifolds modelled on the $(2n-1)$-dimensional Heisenberg group.

Let us take a generic hyperplane $H \subset \mathbb{P}^n$. The intersection $H \cap \bar{V}$ avoids the cusps, so it is contained in $V$. We can apply a noncompact version of the Lefschetz theorem, Theorem 8.8 below, which says that in this case the inclusion $\bar{V} \cap H \hookrightarrow V$ induces isomorphisms

$$\pi_i(V \cap H) \cong \pi_i(V)$$

for $i < n - 1$. In the case $i = 1$, we obtain:

COROLLARY 8.4. *If $n \geq 3$, then $\Gamma$ is a Kähler group.*

REMARK 8.5. The quasi–projective manifold $V$ is an Eilenberg–Mac Lane space, but $V \cap H$ must have non–trivial higher homotopy groups.

Actually, in this situation one can characterise the lattices which are Kähler:

THEOREM 8.6. *Let $\Gamma \subset SU(1,n)$ be a torsion–free, non–cocompact lattice. Then $\Gamma$ is Kähler if and only if $n \geq 3$.*

PROOF. If $n \geq 3$, the same argument as in the above Corollary shows that $\Gamma$ is Kähler. It remains to study the cases $n = 1, 2$.

If $n = 1$, then $\Gamma$ is free, so it cannot be Kähler.

Consider now the case $n = 2$. Suppose $\Gamma \cong \pi_1(M)$ with $M$ a compact Kähler manifold. Then there exists a continuous map

$$f \colon M \longrightarrow \Gamma \backslash B^2$$

inducing the isomorphism of fundamental groups. By the Eells–Sampson Theorem 5.8 we can take $f$ to be harmonic. There are two possibilities:

**1.** rank $f > 2$. By an elaboration of Theorem 6.13 proved in Chapter 6, $f$ is holomorphic, see Remark 6.16. The algebraic fact which makes $f$ holomorphic is that, if $\mathfrak{a} \subset \mathfrak{p}^{\mathbb{C}}$ is Abelian and of dimension greater than one, then $\mathfrak{a} \subset \mathfrak{p}^{1,0}$, see Theorem (6.7) of [**121**], or Theorem (3.5) of [**25**]. Thus $f$ satisfies the Cauchy–Riemann equations. But this is impossible, because $\dim M \geq \dim f(M) = \dim \Gamma \backslash B^2$; but $M$ is compact and $\Gamma \backslash B^2$ is not. One way to see the contradiction is that if $p$ is an interior point of $f(M)$, then $f^{-1}(p)$ is not homologous to zero, since it is a closed analytic subvariety of $M$. But if $q \in (\Gamma \backslash B^2) \setminus f(M)$, then $f^{-1}(q)$ is empty, hence homologically trivial. This contradicts the fact that the homology class of the pre-image of a generic point in the target is independent of the point (provided the target is a connected manifold).

**2.** rank $f \leq 2$. By the factorisation Theorem 6.21, the harmonic map $f$ factors through a map $h \colon M \to S$, with $S = S^1$ or a compact Riemann surface. Again, this is impossible because $f$ induces an isomorphism of fundamental groups and $\Gamma$ has cohomological dimension 3. □

This construction of a projective variety with the same fundamental group as a quasi-projective one containing it can be extended to other locally symmetric situations by using the Satake, Piatetski–Shapiro, Bailey–Borel compactification $\bar{V}$ of $V = \Gamma \backslash D$, where $D$ is a bounded symmetric domain and $\Gamma$ is a non-cocompact arithmetic group of automorphisms of $D$. The properties we need of this compactification are:

1. $\bar{V}$ is projective.
2. $\bar{V} \setminus V$ is a union of varieties $\Gamma' \backslash D'$ where $D'$ are boundary components of $D$.

Next, the Lefschetz Hyperplane Theorem holds for smooth open varieties (see Theorem 8.8 below). Thus it is clear that the above argument for the non-compact ball quotients generalises in the following way:

THEOREM 8.7. *Let $V$ be a smooth quasi-projective variety. If there exists a projective (possibly singular) compactification $\bar{V}$ of $V$ such that the codimension of the complement $\bar{V} \setminus V$ in $\bar{V}$ is at least three, then there is a smooth projective hyperplane section $M \subset \bar{V}$ such that $M \subset V$ and the inclusion induces an isomorphism of fundamental groups $\pi_1(M) \cong \pi_1(V)$.*

Combining this observation with the properties of the Bailey–Borel compactification, it only remains to check the codimensions of the boundary components. One finds that, for irreducible bounded symmetric domains, there exist such projective compactifications $\bar{V}$ with codim $\bar{V} \setminus V \geq 3$ except in the cases when $D$ is the complex ball of dimension 1 or 2 or the dimension two Siegel upper half-plane $\mathcal{H}_2$.

## 3. A Lefschetz theorem for smooth open varieties

The following Theorem contains all versions of the Lefschetz hyperplane theorem that we shall need in our applications.

THEOREM 8.8 (Goresky–MacPherson [53]). *Let $V$ be a smooth complex algebraic variety and $f\colon V \to \mathbb{P}^N$ a holomorphic map with finite fibers. If $L \subset \mathbb{P}^N$ is a generic linear subspace, then*
$$\pi_i(V, f^{-1}(L)) = 0$$
*for $i \leq \dim_\mathbb{C} f^{-1}(L)$.*

We will not prove this result here, but refer to [53]. But we will prove a very simple special case that actually suffices for the first example in the previous Section and for the main example still to come. See [132] for more details.

EXAMPLE 8.9. Let $V \subset \bar{V} = V \cup \{P\} \subset \mathbb{P}^N$ be a pair of varieties, with $V$ smooth and $\bar{V}$ complete. In other words, suppose that $V$ can be compactified by adding a single point $P$. Let $H \subset \mathbb{P}^N$ be a generic hyperplane. Then $V$ is homotopy equivalent to
$$(V \cap H) \cup \{\text{ cells of } \dim \geq n = \dim_\mathbb{C} V\} \ .$$

Let us sketch the argument, which is a small modification of an argument going back to Thom and Bott [19]. We aim to build a suitable Morse function. Take coordinates in $\mathbb{P}^n$ so that $H$ is the hyperplane at infinity and the singular point is $P = 0$. Consider the function
$$\begin{aligned} f : \mathbb{P}^N &\longrightarrow \mathbb{R} \\ x &\longmapsto d^2(H, x) \ , \end{aligned}$$
where $d$ is the distance in the Fubini–Study metric. The level sets of $f$ in the affine open set $\mathbb{C}^N = \mathbb{P}^N \setminus H$ are balls centreed at the origin. Let $Q$ be a critical point of $f$ restricted to $V \setminus H$. The manifold $V$ is pseudo-convex, so the Laplacian of the restriction of $f$ to any curve through $Q$ is negative. Therefore, by taking curves along linearly independent directions we see that the Hessian of $f$ has at least $n$ negative eigenvalues, so $\mathrm{index}\,(Q) \geq n$, and our claim follows.

REMARK 8.10. The only difference between this argument and that in [19] is that we do not use the maximum of $f$, which is the singular point of $\bar{V}$. If we added this singular point, then we would not be adding a cell, but rather the cone on the link of $P$ in $\bar{V}$, which may have non-trivial topology. For example, in the case of $V = \Gamma\backslash B^n$ above, we would be adding the cone on a nilmanifold, thus changing the fundamental group.

REMARK 8.11. In the application to ball quotients we actually need to consider finitely many cusps. This can be handled in the same way, because one can re-embed $V$ in $\mathbb{P}^N$ in such a manner that all the cusps go to the same point, and then one is in the situation just discussed.

## 4. The general construction

We now apply the theorem of Goresky and MacPherson, Theorem 8.8, to the construction of further examples of Kähler groups. In fact, all these further examples are based on one common construction:

Let $A$ and $B$ be projective $n$–dimensional manifolds and $\alpha\colon A \to \mathbb{P}^n$ and $\beta\colon B \to \mathbb{P}^n$ be finite maps. Consider the open set $U \subset \mathbb{P}^{2n+1}$ given by the complement of a generic pair of $n$–dimensional subspaces. Through every point in $U$ there is a unique complex line cutting both linear $n$–spaces. Thus there is a fibration

$$\begin{array}{ccc} \mathbb{C}^* & \longrightarrow & U \\ & & \downarrow \\ & & \mathbb{P}^n \times \mathbb{P}^n \end{array}$$

sending every point in $U$ to the intersections of its line with the linear spaces. Define now $V$ as the pullback in the following diagram:

$$\begin{array}{ccc} V & \longrightarrow & U \\ \downarrow & & \downarrow \\ A \times B & \xrightarrow{\alpha \times \beta} & \mathbb{P}^n \times \mathbb{P}^n \end{array}$$

The composition $V \to U \hookrightarrow \mathbb{P}^{2n+1}$ satisfies the finiteness hypothesis of the Goresky–MacPherson Theorem 8.8.

Take now a new generic linear $n$–space $\Lambda \subset \mathbb{P}^{2n+1}$. In particular, $\Lambda$ is contained in $U$. Denote $M = f^{-1}(\Lambda) \subset V$. This is a projective smooth manifold, and, by Theorem 8.8, if $n \geq 2$ then the inclusion induces an isomorphism of fundamental groups $\pi_1(M) \cong \pi_1(V)$, so $\pi_1(V)$ is a Kähler group.

By its definition, the variety $V$ is a $\mathbb{C}^*$–bundle over $A \times B$. Therefore there is an exact sequence

$$\pi_2(A \times B) \longrightarrow \mathbb{Z} \longrightarrow \pi_1(V) \longrightarrow \pi_1(A \times B) \longrightarrow 1\,.$$

In particular, if $\pi_2(A \times B) = 0$, then the fundamental group of $V$ is a central extension

$$1 \longrightarrow \mathbb{Z} \longrightarrow \pi_1(V) \longrightarrow \pi_1(A \times B) \longrightarrow 1\,.$$

This central extension has characteristic class $\alpha^*\omega - \beta^*\omega \in H^2(\pi_1(A \times B), \mathbb{Z})$, where $\omega \in H^2(\mathbb{P}^n, \mathbb{Z})$ is the characteristic class of the tautological line sub-bundle $L$ over $\mathbb{P}^n$.

To use this construction to produce Kähler groups, one has to find appropriate maps $\alpha$ and $\beta$.

### 4.1. Nilpotent Kähler groups.

EXAMPLE 8.12. (Sommese–Van de Ven [124], Campana [23]) Take $A$ and $B$ to be $n$–dimensional Abelian varieties. Then $\pi_1(A \times B) \cong \mathbb{Z}^{4n}$, so $\pi_1(V)$ is a central extension

$$0 \longrightarrow \mathbb{Z} \longrightarrow \pi_1(V) \longrightarrow \mathbb{Z}^{4n} \longrightarrow 0\,.$$

Its characteristic class turns out to be a symplectic form, and therefore $\pi_1(V)$ is a lattice in the $(4n+1)$–dimensional Heisenberg group $H_{4n+1} \subset GL(2n+2, \mathbb{R})$.

If $\dim A = n$, $\dim B = n+1$, a parallel argument yields lattices in $H_{4n+3} \subset GL(2n+3, \mathbb{R})$ as Kähler groups, provided that $n \geq 2$.

This construction proves the "if" part of the following Theorem concerning the $(2k+1)$-dimensional Heisenberg group $H_{2k+1} \subset GL(k+2, \mathbb{R})$:

THEOREM 8.13 (Carlson–Toledo [26], Campana [23]). *A lattice* $\Gamma \subset H_{2k+1}$ *is Kähler if and only if* $k \geq 4$.

## 5. Non–residually finite Kähler groups

Using the construction in the previous Section, we will now give a number of examples of non–residually finite Kähler groups. As was mentioned before, they all depend on properties of lattices in non–linear Lie groups.

Let $G$ be a linear Lie group such that $G/K$ is an irreducible Hermitian symmetric space. This implies that $K = S^1 \times K_0$, with $\pi_1(K_0)$ finite, so there is a circle bundle

$$S^1 \longrightarrow G/K_0 \\ \downarrow \\ G/K$$

whose characteristic class is the Kähler class of $G/K$.

Take the universal cover $(G/K_0)^\sim = \hat{G}/K_0$, where $\hat{G}$ is the covering of $G$ corresponding to the inclusion $\pi_1(S^1) \subset \pi_1(S^1) \times \pi_1(K_0) \cong \pi_1(G)$ and $K_0$ is identified with its unique lift as a subgroup of $\hat{G}$. The group $\hat{G}$ is a non–linear Lie group, because it is semi–simple and its centre contains $\mathbb{Z}$. A particular example of such a group arises by taking $G = SO_0(2,n)$ (the identity component of $SO(2,n)$). In this case we denote $\hat{G}$ simply by $\widehat{SO}(2,n)$. This group will be crucial to all our remaining examples because of the following remarkable Theorem.

THEOREM 8.14 (Raghunathan [**105**]). *Let $\Gamma$ be a cocompact lattice in $SO_0(2,n)$, $n$ odd and $n \geq 3$. Define $\hat{\Gamma} \subset \widehat{SO}(2,n)$ to be the pre–image of $\Gamma$, which is a lattice in $\widehat{SO}(2,n)$ and is an extension*

$$1 \longrightarrow \mathbb{Z} \longrightarrow \hat{\Gamma} \longrightarrow \Gamma \longrightarrow 1.$$

*If $H \subset \hat{\Gamma}$ is a subgroup of finite index, then $H \supset 8\mathbb{Z}$.*

The proof rests on the congruence subgroup property for $\Gamma$, which is the group of units of an integral quadratic form. The fact that $\mathbb{Z}$ is Abelian also plays an essential role, as shown by Theorem A.14 of Appendix A.

This result will allow us to construct examples of non–residually finite Kähler groups. Here is a simple class of examples (presented in a slightly different way in [**30**]).

EXAMPLE 8.15. (Catanese–Kollár [**30**], Nori (unpublished)) Take a finite holomorphic map $\alpha \colon A = \Gamma \backslash G/K \to \mathbb{P}^n$, where $G = SO_0(2,n)$ with $n \geq 3$, and $\Gamma$ is a discrete, torsion–free, cocompact subgroup of $G$. Further, let $\beta \colon \mathbb{P}^n \to \mathbb{P}^n$ be a map of odd degree $d = \deg \beta$. Applying the above construction in this case, we get a pullback of $\mathbb{C}^*$-bundles

$$\begin{array}{ccc} V & \longrightarrow & U \\ \downarrow & & \downarrow \\ A \times B & \stackrel{\alpha \times \beta}{\longrightarrow} & \mathbb{P}^n \times \mathbb{P}^n \end{array}$$

and the corresponding homotopy exact sequence is

$$\pi_2(A \times B) \cong \mathbb{Z} \stackrel{\cdot d}{\longrightarrow} \mathbb{Z} \longrightarrow \pi_1(V) \longrightarrow \Gamma \longrightarrow 1.$$

Thus we have obtained a central extension

$$0 \longrightarrow \mathbb{Z}/d\mathbb{Z} \longrightarrow \pi_1(V) \longrightarrow \Gamma \longrightarrow 1.$$

The Kähler group $\pi_1(V)$ is a lattice in the quotient $\hat{G}/d\mathbb{Z}$. Since $d$ is odd, Raghunathan's theorem implies that the intersection of the subgroups of finite index of $\pi_1(V)$ is $\mathbb{Z}/d\mathbb{Z}$. In particular, $\pi_1(V)$ is not residually finite.

EXAMPLE 8.16. (Toledo) Take $\alpha\colon A = \Gamma\backslash G/K \to \mathbb{P}^n$ exactly as above, but take $B$ to be any aspherical variety and $\beta\colon B \to \mathbb{P}^n$ a finite map. Proceeding in the same way we obtain a variety $V$ whose fundamental group is a central extension

$$0 \longrightarrow \mathbb{Z} \longrightarrow \pi_1(V) \longrightarrow \Gamma \longrightarrow 1\,.$$

Again, $\pi_1(V)$ is not residually finite because it contains the group in Raghunathan's theorem as a subgroup. In this example, the intersection of subgroups of finite index is infinite cyclic, while in the Nori and Catanese–Kollár examples it is a finite cyclic group.

It is clear from the above examples that the following Theorem holds when $C$ has no even torsion:

THEOREM 8.17. *Let $C$ be a finitely generated Abelian group. Then there exists a smooth projective variety $M$ with the property that the intersection of all subgroups of finite index in $\pi_1(M)$ is isomorphic to $C$.*

In order to prove the Theorem for even torsion, one needs later refinements of Raghunathan's theorem, due to Prasad, to the effect that the intersection of all subgroups of finite index is precisely $2\mathbb{Z}$. Alternatively, one can use Deligne's Example 8.3, where the intersection of all subgroups of finite index is precisely $2\mathbb{Z}$. For the variety $A$ in Example 8.15 take a 2–connected projective hyperplane section of a Siegel upper half plane modulo the integral symplectic group, as explained in [**30**]. If $d$ is an even integer and we take $\beta\colon \mathbb{P}^n \to \mathbb{P}^n$ to be a map of degree $2d$, then the construction of Example 8.15 gives a Kähler group which is a central extension

$$0 \longrightarrow \mathbb{Z}/2d\mathbb{Z} \longrightarrow \pi_1(V) \longrightarrow Sp(2n,\mathbb{Z}) \longrightarrow 1$$

and with intersection of all subgroups of finite index isomorphic to $\mathbb{Z}/d\mathbb{Z}$.

From Miller's Theorem A.14, it is easy to see that if the intersection of all subgroups of finite index of a group $\Gamma$ is itself residually finite and finitely generated, then it must be central, and in particular Abelian (cf. also Proposition (6.4.1) of [**80**]). Thus the above Theorem says that all intersections of subgoups of finite index which are residually finite and finitely generated can be realised by Kähler groups. As of this writing there is no known example of a Kähler group with the property that the intersection of its subgroups of finite index is itself not residually finite[1]. But, to end this Chapter, we now present the original example of a non–residually finite Kähler group due to Toledo [**133**], which has the property that this intersection is a free group of infinite rank.

Let us consider the symmetric space for $SO_0(2,n)$

$$X_n = SO_0(2,n)/(SO(2) \times SO(n))\,.$$

---

[1]Some attempts and interesting ideas in this direction can be found in [**13**].

This space parametrises the set of 2–dimensional positive definite planes in $\mathbb{R}^{n+2}$ for the bilinear form

$$b(x,y) = x^t \begin{pmatrix} 1 & & & & \\ & 1 & & & \\ & & -1 & & \\ & & & \ddots & \\ & & & & -1 \end{pmatrix} y \ .$$

Moreover, it admits a complex manifold structure: Let $x, y$ be a $b$–orthogonal basis for a positive 2–dimensional plane in $\mathbb{R}^{n+2}$. There is a well–defined map

$$\nu \colon X_n \longrightarrow \mathbb{C}P^{n+1}.$$
$$\langle x, y \rangle \longmapsto [z] = [x + iy] \ .$$

The quadratic form $q(x) = b(x,x)$ on $\mathbb{R}^{n+2}$ extends to a quadratic form on $\mathbb{C}^{n+2}$ as $q(z) = q(x) - q(y) + 2ib(x,y)$ for $z = x + iy$. The map $\nu$ is an inclusion, and one reaches the characterisation

$$\nu(X_n) = \{[z] \in \mathbb{C}P^{n+1} \mid q(z) = 0, \ b(z, \bar{z}) > 0\} \ .$$

We will henceforth identify $X_n$ and $\nu(X_n)$. It is thus a complex manifold contained in a quadric hypersurface in $\mathbb{C}P^{n+1}$.

Take now the tautological line bundle $S \subset X_n \times \mathbb{C}^{n+2}$ over $X_n$. It has positive degree, and its Chern class $c_1(S)$ is the Kähler form of $X_n$. The canonical bundle of $X_n$ is

$$K_{X_n} \cong S^{\otimes n} \ ,$$

and this isomorphism is compatible with the $SO_0(2,n)$-action on both sides. Let us fix a vector $v \in \mathbb{R}^{n+2}$ such that $q(x) < 0$. The restricted quadratic form $q_{|v^\perp}$ has signature $(2, n-1)$. The isotropy group of $v$ in $SO_0(2,n)$ is isomorphic to $SO_0(2, n-1)$, and we will denote it by $SO_0(2, n-1)_v$. The isomorphism

$$X_{n-1} \cong \{z \in X_n \mid b(z, v) = 0\}$$

allows us to identify $X_{n-1}$ with a complex submanifold of $X_n$. As this embedding depends on the choice of vector $v$, it will be denoted as $X^v_{n-1} \subset X_n$. There is a section of the dual bundle $S^*$ given by

$$z \longmapsto b(z, v)$$

This section vanishes exactly on $X^v_{n-1}$.

The next step is to choose a torsion–free cocompact lattice $\Gamma \subset SO_0(2,n)$ and a negative vector $v \in \mathbb{R}^{n+2}$ such that the intersection

$$\Gamma_v = SO_0(2, n-1)_v \cap \Gamma$$

is a lattice, and the map

$$D = \Gamma_v \backslash X^v_{n-1} \hookrightarrow \Gamma \backslash X_n = M$$

is an embedding with image a smooth divisor in $M$. Classical arguments, explained in section 4.3 of Borel's 1963 paper [14] on a much more general construction of lattices, show that one can do this by taking $q$ with coefficients in suitable real quadratic fields, such as

$$q(x) = \sqrt{2}(x_1^2 + x_2^2) - (x_3^2 + \cdots + x_{n+2}^2) \ .$$

In such a case, one can choose $\Gamma$ as a torsion–free subgroup of finite index of $SO(q, \mathbb{Z}[\sqrt{2}])$. The inclusion $D \hookrightarrow M$ is then an embedding because $D$ is a connected component of the fixed point set of an isometry of $M = \Gamma \backslash X_n$, namely the isometry arising from reflection on $v^\perp$.

Thus we have obtained a divisor $D \subset M$, with associated line bundle

$$\mathcal{O}(D) = \{s \in \mathcal{O}_M \mid s = 0 \text{ exactly on } D\} .$$

For a small neighbourhood $D \subset U \subset M$ there is an isomorphism

$$\mathcal{O}(D)_{|U} \cong S^*_{|U} .$$

Thus the normal bundle of $D$ in $M$ is $S^*_{|D}$. Consequently, the sheaf $K_M(nD) = K_M \otimes (\mathcal{O}(D))^{\otimes n}$ is trivial on $U$, so it descends to a line bundle on $M/D = M/\sim$, where $x \sim y$ if $x, y \in D$.

LEMMA 8.18. *For some $k > 0$ the line bundle $K_M(nD)^{\otimes k}$ embeds $M/D$ in a projective space.*

PROOF. We must examine the meromorphic map given by

$$M \dashrightarrow \mathbb{P} = \mathbb{P}(H^0(M, K_M(nD)^{\otimes k})^*)$$
$$p \longmapsto \{s \in H^0(M, K_M(nD)^{\otimes k}) \mid s(p) = 0\}$$

*Claim 1*: This map is actually defined for all $p \in M$.
*Claim 2*: This map induces an embedding of $M \setminus D$ in $\mathbb{P}$.

The claims can be proved by standard techniques in algebraic geometry. □

This Lemma induces a projective variety structure on $M/D$, which is the main technical point of our construction of non–residually finite Kähler groups. We are now ready to present the example.

Let $H \subset \mathbb{P}$ be a generic hyperplane. The intersection

$$Z = H \cap (M/D) \subset H \cap (M \setminus D)$$

is a projective manifold. By the Lefschetz Theorem 8.8 for smooth open varieties, provided that $\dim \mathbb{P} \geq 3$, the inclusion yields an isomorphism of fundamental groups

$$\pi_1(Z) \xrightarrow{\cong} \pi_1(M \setminus D) .$$

Therefore, $\Phi = \pi_1(M \setminus D)$ is the fundamental group of a projective variety. It remains to study its pro–finite completion. As the codimension of $D$ in $M$ is 2, the map $\pi_1(M \setminus D) \to \pi_1(M)$ is surjective, and there is an exact sequence

$$1 \longrightarrow K \longrightarrow \Phi \longrightarrow \Gamma \longrightarrow 1 .$$

Let $N$ be a tubular neighbourhood of $D$ in $M$. Topologically we may view it as the associated disk bundle of the normal bundle of $D$, $S^*_{|D}$. Thus $\partial N$ is the circle bundle of $S^*_{|D}$:

$$SO(2) = S^1 \longrightarrow \Phi_v \backslash \widehat{SO}(2, n-1)/SO(n-1)$$
$$\downarrow$$
$$\Gamma \backslash SO_0(2, n-1)/(\check{S}O(2) \times SO(n-1))$$

Furthermore, there is an injective map of homotopy long exact sequences of the circle and disk bundles:

$$\begin{array}{ccccc}
\mathbb{Z} & = & \mathbb{Z} & \stackrel{f}{\hookrightarrow} & K \\
\downarrow & & \downarrow & & \downarrow \\
\Phi_v & = & \pi_1(\partial N) & \stackrel{g}{\to} & \pi_1(M \setminus D) \\
\downarrow & & \downarrow & & \downarrow \\
\Gamma_v & = & \pi_1(D) & \stackrel{h}{\hookrightarrow} & \pi_1(M)
\end{array}$$

The maps $f$ and $h$ are injective, the latter because $D$ is a totally geodesic submanifold. Thus $g$ is injective. The group $\Phi_v$ is non–residually finite by Theorem 8.14. Therefore, as $\Phi_v \subset \Phi = \pi_1(M \setminus D)$, the group $\Phi$ cannot be residually finite, which is what we wanted to show.

Moreover, if for each $w$ in the orbit $\Gamma v$ of $v$, we let $z_w \in \Phi_w$ be the generator of the centre, then $z_w^8 \in \bigcap_{\substack{H \subset \Phi \\ f.i.}} H$. Thus, if $\langle x_a \mid a \in A \rangle$ denotes the subgroup of $\Phi$ generated by a subset of $\Phi$, indexed by a set $A$, we see that

$$\langle z_w^8 \mid w \in \Gamma v \rangle \subset \bigcap_{\substack{H \subset \Phi \\ f.i.}} H \subset \langle z_w \mid w \in \Gamma v \rangle .$$

Finally, it can be shown (and is geometrically quite evident from the fact that $\langle z_w \mid w \in \Gamma v \rangle$ is the fundamental group of the complement in $X_n$ of the union of an infinite disjoint family of totally geodesic real codimension two subspaces) that $\langle z_w \mid w \in \Gamma v \rangle$ as well as $\langle z_w^8 \mid w \in \Gamma v \rangle$ are free groups of infinite rank. From this it follows that $\bigcap_{\substack{H \subset \Phi \\ f.i.}} H$ is itself a free group of infinite rank, as was claimed above.

More details of this example can be found in [133].

# APPENDIX A

# Pro group theory

This Appendix collects the basic constructions and properties of projective completions of groups used in this book.

## 1. Definitions of group completions

Categorically, group completions are defined as left adjoint functors of inclusions.

We denote by $\mathcal{P}$ a subcategory of the category of groups $\mathcal{G}r$ and by $i\colon \mathcal{P} \hookrightarrow \mathcal{G}r$ its inclusion functor. For example, $\mathcal{P}$ could be the category of finite or of nilpotent groups, with morphisms their group homomorphisms. We will denote by $Hom_{\mathcal{P}}$ the morphisms of the category $\mathcal{P}$.

The defining property of the left adjoint functor $\cdot^{\mathcal{P}}\colon \mathcal{G}r \to \mathcal{P}$, if it exists, is that for every pair of objects $\Gamma \in \mathcal{G}r$, $\Delta \in \mathcal{P}$, there is a bijection

$$(12) \qquad Hom_{\mathcal{G}r}(\Gamma, i(\Delta)) \overset{\cong}{\longleftrightarrow} Hom_{\mathcal{P}}(\Gamma^{\mathcal{P}}, \Delta),$$

which is natural in both arguments.

DEFINITION A.1. Let $\mathcal{G}r$ be the category of groups, and $\mathcal{P} \subset \mathcal{G}r$ a subcategory such that the inclusion functor $i\colon \mathcal{P} \hookrightarrow \mathcal{G}r$ has a left adjoint functor

$$\cdot^{\mathcal{P}}\colon \mathcal{G}r \longrightarrow \mathcal{P}.$$

The $\mathcal{P}$-*completion of a group* $\Gamma$ is the homomorphism

$$\eta^{\mathcal{P}}\colon \Gamma \longrightarrow \Gamma^{\mathcal{P}},$$

which is the element of $Hom_{\mathcal{G}r}(\Gamma, i(\Gamma^{\mathcal{P}}))$ corresponding to the identity of $\Gamma^{\mathcal{P}}$ under the adjointness bijection (12).

There is an equivalent definition of the $\mathcal{P}$-completion in terms of a universal property:

DEFINITION A.2. Let $\Gamma$ be a group. The $\mathcal{P}$-*completion* of $\Gamma$ is a group homomorphism

$$\eta^{\mathcal{P}}\colon \Gamma \longrightarrow \Gamma^{\mathcal{P}},$$

where $\Gamma^{\mathcal{P}}$ is an object of $\mathcal{P}$ having the universal property for homomorphisms from $\Gamma$ to groups in $\mathcal{P}$.

In other words, $\eta^{\mathcal{P}}$ has the following property: every group homomorphism $f\colon \Gamma \longrightarrow \Delta$, where $\Delta$ is in $\mathcal{P}$, factors uniquely through $\eta^{\mathcal{P}}$ to give a commutative diagram

$$\begin{array}{ccc} \Gamma & \overset{\eta^{\mathcal{P}}}{\longrightarrow} & \Gamma^{\mathcal{P}} \\ {\scriptstyle f}\downarrow & \swarrow {\scriptstyle f^{\mathcal{P}}} & \\ \Delta & & \end{array}$$

The universal property implies that the universal object it defines is functorial in $\Gamma$ and is unique up to canonical isomorphism. However, a universal object with the postulated property may not exist, as in the following example:

EXAMPLE A.3. As $\mathbb{Z}$ has finite quotients of arbitrarily large order, there cannot exist any finite group through which all the quotient homomorphisms factor.

By Definition A.1, the existence of the $\mathcal{P}$-completion is equivalent to the existence of a left adjoint for the inclusion functor $i\colon \mathcal{P} \hookrightarrow \mathcal{G}r$. There is a categorical sufficient condition for this. In the situation at hand, the condition is very simple:

PROPOSITION A.4. *If $\mathcal{P}$ is a subcategory of the category of groups closed under projective limits, then the inclusion functor $i\colon \mathcal{P} \to \mathcal{G}r$ has a left adjoint.*

PROOF. Using the universal property, rather than the categorical definition, one constructs the objects $\Gamma^{\mathcal{P}}$ for $\mathcal{P}$ closed under projective limits as a projective limit of homomorphisms $\Gamma \to \Delta \in \mathcal{P}$. □

The fact that not every subcategory defines a completion is remedied by considering the projective closure of a category, e.g. *pro*–finite or *pro*–nilpotent groups and completions. Nevertheless, the $\mathcal{P}$-completion $\eta^{\mathcal{P}}\colon \Gamma \to \Gamma^{\mathcal{P}}$ is determined by the homomorphisms from $\Gamma$ to any cofinal set of objects in $\mathcal{P}$, so the pro–finite or pro–nilpotent completions are determined by their universal properties with respect to homomorphisms of $\Gamma$ into finite or nilpotent groups. In fact, we may and do define the $\mathcal{P}$-completion of a group for *any* subcategory $\mathcal{P}$ of groups as the left adjoint of the inclusion functor of its projective closure in the category of all groups.

Here are some examples of $\mathcal{P}$-completions:

EXAMPLE A.5. It is customary to denote the *pro–finite completion* $\eta^{\text{finite}}\colon \Gamma \to \Gamma^{\text{finite}}$ by
$$\hat{\eta}\colon \Gamma \longrightarrow \hat{\Gamma}.$$
When $\Gamma$ is the fundamental group of an algebraic variety, $\hat{\Gamma}$ is its *algebraic fundamental group*.

EXAMPLE A.6. The *pro-l-finite completion* is the completion with respect to the subcategory of pro–$l$–groups, i.e., projective limits of finite groups of order $l^n$, where $l$ is any prime. It is denoted as
$$\eta_l^{\wedge}\colon \Gamma \longrightarrow \Gamma_l^{\wedge}$$
and may be of interest for a motivic study of fundamental groups.

EXAMPLE A.7. The *nilpotent completion*
$$\eta^{nilp}\colon \Gamma \longrightarrow \Gamma^{nilp}$$
corresponds to taking for $\mathcal{P}$ the category $Nilp$ of pro–nilpotent groups.

EXAMPLE A.8. The *torsion–free nilpotent completion* corresponds to the case when $\mathcal{P}$ is the projective completion $Nilp_0$ of the full subcategory of torsion–free nilpotent groups and will be denoted by[1]
$$\eta_0^{nilp}\colon \Gamma \longrightarrow \Gamma_0^{nilp}.$$

---

[1] This notation conflicts with [**23**], where our $\Gamma_0^{nilp}$ is denoted $\Gamma^{nilp}$, which is more naturally reserved for the previous example.

EXAMPLE A.9. The *k–unipotent completion* is obtained by taking for $\mathcal{P}$ the category $U_k$ of projective limits of $k$–unipotent groups, where $k$ is any field of characteristic zero. This is denoted by

$$\eta \otimes k : \Gamma \longrightarrow \Gamma \otimes k \ .$$

When $k = \mathbb{R}$ and $\Gamma$ is the fundamental group of a smooth manifold, $\Gamma \otimes \mathbb{R}$ is its *de Rham fundamental group*.

EXAMPLE A.10. The *algebraic hull*[2] of $\Gamma$

$$\eta^{alg} : \Gamma \longrightarrow \Gamma^{alg}$$

is obtained by taking for $\mathcal{P}$ the projective completion of the category of affine algebraic groups over $\mathbb{C}$.

A very desirable property of a completion functor $\cdot^{\mathcal{P}}$ is *idempotence*, i.e., $\cdot^{\mathcal{P}} \circ i \circ \cdot^{\mathcal{P}} = \cdot^{\mathcal{P}}$, or the slightly stronger property that $\cdot^{\mathcal{P}} \circ i = Id_{\mathcal{P}}$, that is, $\Gamma^{\mathcal{P}} \cong \Gamma$ naturally for all $\Gamma \in \mathcal{P}$. The latter admits a neat description in our case:

LEMMA A.11. *Let $\mathcal{P}$ be a group subcategory closed under projective limits. Then $\eta^{\mathcal{P}}$ is a natural isomorphism $\Gamma^{\mathcal{P}} \cong \Gamma$ for all $\Gamma \in \mathcal{P}$ if and only if $\mathcal{P} \subset \mathcal{G}r$ is a full subcategory, i.e., $Hom_{\mathcal{P}}(\Gamma, \Delta) = Hom_{\mathcal{G}r}(\Gamma, \Delta)$ for all $\Gamma, \Delta \in \mathcal{P}$.*

PROOF. The completion $\cdot^{\mathcal{P}}$ exists, and by its definition we have

$$Hom_{\mathcal{G}r}(\Gamma, i(\Delta)) \xleftrightarrow{\cong} Hom_{\mathcal{P}}(\Gamma^{\mathcal{P}}, \Delta)$$

for every $\Gamma \in \mathcal{P}$.

If $\Gamma^{\mathcal{P}} \cong \Gamma$ for all $\Gamma \in \mathcal{P}$, this is the definition of fullness for $\mathcal{P}$. Conversely, if $\mathcal{P}$ is full, setting $\Gamma^{\mathcal{P}} \cong \Gamma$ for all $\Gamma \in \mathcal{P}$, the identity of $\Gamma \in \mathcal{P}$ satisfies the universal property defining $\eta^{\mathcal{P}}$. □

Looking at the above examples, it turns out that the completions A.5–A.8 are idempotent, but the $k$–unipotent completion and the algebraic hull are not, because of algebraicity conditions on the morphisms of $k$–unipotent and affine complex–algebraic groups.

When $\eta^{\mathcal{P}}$ is not an isomorphism, it may still be injective *or* surjective. More generally, it is interesting to study its kernel and cokernel. The case when the completion is injective is of particular importance.

DEFINITION A.12. A group $\Gamma$ is *residually $\mathcal{P}$* if its completion $\eta^{\mathcal{P}} : \Gamma \to \Gamma^{\mathcal{P}}$ is injective.

Being residually $\mathcal{P}$ is equivalent to the following condition on $\Gamma$: for each $g \in \Gamma$ which is not the neutral element there exists a homomorphism $f : \Gamma \to \Delta$ with $\Delta \in \mathcal{P}$ such that $f(g) \neq e$.

EXAMPLE A.13. The group $\mathbb{Z}$, and in general torsion–free Abelian groups, are residually $\mathcal{P}$ in all of the previous examples. The same holds for finite rank free groups, although the proof is less straightforward.

In Chapter 8, we use the following criterion for residual finiteness:

---

[2] Some authors use the notation $\pi_1^{alg}$ for the algebraic fundamental group of Example A.5. This should not be confused with the algebraic hull.

THEOREM A.14 (Miller [**92**]). *Consider an extension of groups*

$$1 \longrightarrow A \longrightarrow B \longrightarrow C \longrightarrow 1$$

*such that*

1. *the groups $A$ and $C$ are residually finite, and*
2. *$A$ is finitely generated and has trivial center.*

*Then $B$ is residually finite.*

We conclude this section with some easy examples of $\mathcal{P}$-completions:

EXAMPLE A.15.  1. $\Gamma = \{1\} \Longrightarrow \hat{\Gamma} = \Gamma_0^{\mathrm{nilp}} = \Gamma \otimes k = \{1\}$,
2. $|\Gamma| < \infty \Longrightarrow \hat{\Gamma} = \Gamma$ and $\Gamma_0^{\mathrm{nilp}} = \Gamma \otimes k = \{1\}$,
3. $\Gamma = \mathbb{Z} \Longrightarrow \hat{\Gamma} = \prod_{p \text{ prime}} \mathbb{Z}_p$, $\Gamma_0^{\mathrm{nilp}} = \Gamma = \mathbb{Z}$ and $\Gamma \otimes k = k$,
4. $\Gamma$ finitely generated Abelian $\Longrightarrow \hat{\Gamma} = (\hat{\mathbb{Z}})^{\mathrm{rank}(\Gamma)} \times \mathrm{Tor}(\Gamma)$, $\Gamma_0^{\mathrm{nilp}} = \Gamma/\mathrm{Tor}$ and $\Gamma \otimes k = \Gamma \otimes_{\mathbb{Z}} k$ (usual tensor product, see Example A.17).

## 2. Nilpotent completions

The nilpotent, torsion–free nilpotent and $k$–unipotent completions described in the above examples are in fact strongly related to each other. They can all be obtained by taking limits of simple projective systems associated with the lower central series. In many cases they can even be computed explicitly.

For any two elements $a, b \in \Gamma$, their commutator is denoted by $[a, b]$. The commutator of two subgroups $G, H \subset \Gamma$ is defined to be the subgroup of $\Gamma$ generated by the commutators $[a, b]$ with $a \in G, b \in H$. The *lower central series* of a group $\Gamma$ is defined recursively by

$$\Gamma_1 = \Gamma, \qquad \Gamma_{n+1} = [\Gamma_n, \Gamma].$$

We can naturally assign to $\Gamma$ a tower of nilpotent quotients

(13) $$\ldots \longrightarrow \Gamma/\Gamma_3 \longrightarrow \Gamma/\Gamma_2.$$

It is immediate to check that homomorphisms from $\Gamma$ to nilpotent groups of nilpotency class $n$ factor through $\Gamma/\Gamma_{n+1}$. Thus, the tower (13) is cofinal for homomorphisms from $\Gamma$ to all nilpotent groups. Therefore, its projective limit is the *nilpotent completion*:

$$\Gamma^{nilp} = \varprojlim \Gamma/\Gamma_n$$

and the homomorphism $\eta^{nilp}$ is induced by the projections $\Gamma \to \Gamma/\Gamma_n$.

The torsion–free nilpotent completion can be computed in a similar manner. For a nilpotent group $H$, the torsion elements form a normal subgroup, which is finite if $H$ is finitely generated. Thus, by dividing by this subgroup, we can naturally associate to $H$ a torsion–free nilpotent group $H/_{Tor}$ and to every group $\Gamma$ a tower of torsion–free nilpotent groups

(14) $$\ldots \longrightarrow (\Gamma/\Gamma_3)/_{Tor} \longrightarrow (\Gamma/\Gamma_2)/_{Tor}.$$

As in the case of the nilpotent completion, the torsion–free nilpotent completion of $\Gamma$ is the projective limit of this tower, together with the homomorphism induced by the projections of $\Gamma$ onto the finite stages of the tower.

The $k$–unipotent completion also admits a characterisation in terms of the lower central series, but the details are more involved. In the case of *finitely presentable* groups, there is an alternative description in terms of the group algebra.

Let $\Gamma$ be a finitely presentable group, and $k$ a field of characteristic zero. The *group algebra* $k\Gamma$ has an augmentation $k$-algebra homomorphism

$$\varepsilon\colon k\Gamma \longrightarrow k$$
$$g \in \Gamma \longmapsto 1 \ .$$

Denote by $J = \ker \varepsilon$ the *augmentation ideal*, and by $\widehat{k\Gamma} = \varprojlim k\Gamma/J^n$ the $J$-adic completion of the group algebra[3]. The *completed group algebra* $\widehat{k\Gamma}$ has a coproduct defined by

$$\Delta(g) = g \otimes g$$

for all $g \in \Gamma$, which makes it into a complete Hopf algebra. As such, it has two subsets of distinguished elements:

1. The *group-like elements*

$$\mathcal{G}(\widehat{k\Gamma}) = \{x \in 1 + \hat{J} \mid \Delta(x) = x \otimes x\},$$

which form a group with respect to the multiplication in the algebra.

2. The *primitive elements*

$$\mathcal{P}(\widehat{k\Gamma}) = \{x \in \hat{J} \mid \Delta x = x \otimes 1 + 1 \otimes x\},$$

which form a Lie algebra over $k$ with the bracket defined by commutation in the algebra.

For the proof of the following Proposition, we refer to [**103**]:

PROPOSITION A.16 (Quillen). *For a finitely presentable group $\Gamma$, its $k$–unipotent completion is*

$$\Gamma \otimes k \cong \mathcal{G}(\widehat{k\Gamma}) \ ,$$

*with the morphism $\eta \otimes k$ induced by the natural inclusion $\Gamma \hookrightarrow k\Gamma$.*

EXAMPLE A.17. If $\Gamma$ is an Abelian group, the above construction yields the ordinary tensor product, i.e., $\Gamma \otimes k \cong \Gamma \underset{\mathbb{Z}}{\otimes} k$, which is easily seen to have the universal property for homomorphisms to unipotent groups. This justifies the notation for the $k$–unipotent completions.

For any field $k$ of characteristic zero, there is a well–known equivalence between unipotent groups and nilpotent Lie algebras over $k$. This equivalence is given, for instance, by the Baker–Campbell–Hausdorff formula, and it is compatible with projective limits. This induces a categorical equivalence between pro–$k$–unipotent groups and pro–$k$–nilpotent Lie algebras. It is often convenient to work with the Lie algebra, rather than with the group.

DEFINITION A.18. Let $\Gamma$ be a group. The *$k$-Malcev algebra* of $\Gamma$, denoted by $\mathcal{L}(\Gamma, k)$, is the pro–$k$–nilpotent Lie algebra of the pro–$k$–unipotent group $\Gamma \otimes k$.

When the group $\Gamma$ is finitely presentable, its Malcev algebra can also be characterised in terms of the completed group algebra as the Lie algebra of primitive elements:

$$\mathcal{L}(\Gamma, k) \cong \mathcal{P}(\widehat{k\Gamma}) \ .$$

---

[3]This notation has nothing to do with the pro–finite completion. In particular, $\widehat{k\Gamma}$ and $k\hat{\Gamma}$ are entirely different algebras.

The $k$–Malcev algebra of $\Gamma$ may be presented as a projective limit, analogous to that of the nilpotent completion $\Gamma^{nilp}$,

$$\cdots \to \mathcal{L}_3(\Gamma, k) \to \mathcal{L}_2(\Gamma, k) \to \mathcal{L}_1(\Gamma, k),$$

where $\mathcal{L}_n(\Gamma, k)$ is the $n$–step nilpotent $k$-Malcev algebra of $\Gamma$: it may be defined as the $k$–Malcev algebra of the group $\Gamma/\Gamma_{n+1}$, or alternatively when $\Gamma$ is finitely presentable as the quotient $\mathcal{P}(\widehat{k\Gamma})/\left(\mathcal{P}(\widehat{k\Gamma}) \cap \hat{J}^{n+1}\right)$. Also when $\Gamma$ is finitely presentable, the $n$–step Malcev algebras are finite-dimensional, and $\mathcal{L}_n(\Gamma, k)$ determines the limit algebra $\mathcal{L}(\Gamma, k)$ when $n$ is sufficiently large.

To end this Section, here is a convenient homological characterisation of homomorphisms inducing isomorphisms of $\mathbb{Q}$–unipotent completions:

THEOREM A.19 (Stallings [125]). *Let $f: \Gamma \longrightarrow \Delta$ be a homomorphism between finitely presentable groups with $f_i: H^i(\Gamma, \mathbb{Q}) \longrightarrow H^i(\Delta, \mathbb{Q})$ bijective for $i \leq 1$ and injective for $i = 2$. Then $f \otimes \mathbb{Q}: \Gamma \otimes \mathbb{Q} \longrightarrow \Delta \otimes \mathbb{Q}$ is an isomorphism.*

As we will see in the following section, this theorem is actually valid with coefficients in any field $k$ of characteristic zero.

## 3. Comparison of nilpotent completions

In this Section we compare the different nilpotent completions of a finitely presentable group $\Gamma$. In general, one has the following result:

LEMMA A.20. *If $\mathcal{P}' \subset \mathcal{P} \subset \mathcal{G}r$ are nested subcategories, then there are functorial homomorphisms*

$$\Gamma^{\mathcal{P}} \longrightarrow \Gamma^{\mathcal{P}'}$$

*which, when factorised through $\eta^{\mathcal{P}'}$, yield canonical isomorphisms*

$$(\Gamma^{\mathcal{P}})^{\mathcal{P}'} \xrightarrow{\cong} \Gamma^{\mathcal{P}'}.$$

This is clear, because the homomorphisms of $\Gamma$ to groups in $\mathcal{P}$ contain the homomorphisms to groups in $\mathcal{P}'$. Formally, one can check the required universal property by a diagram chase.

REMARK A.21. It is a consequence of the lemma that, if $\Gamma^{\mathcal{P}} \cong \Delta^{\mathcal{P}}$, then a fortiori $\Gamma^{\mathcal{P}'} \cong \Delta^{\mathcal{P}'}$ for all subcategories $\mathcal{P}' \subset \mathcal{P}$.

This Lemma and Remark apply to the nilpotent completions because of the following inclusions:

$$U_k \subset Nilp_0 \subset Nilp.$$

Some properties of the functorial homomorphisms associated with these inclusions are given by the following Lemmata.

LEMMA A.22. *For every finitely presentable group $\Gamma$, the natural homomorphism $\Gamma^{nilp} \longrightarrow \Gamma_0^{nilp}$ is surjective and has a pro–finite kernel.*

PROOF. The natural homomorphisms between the finite stages of the towers (13) and (14) are surjective and have finite kernels. Passing to the projective limits proves the statement. □

LEMMA A.23. *The natural homomorphism $\Gamma_0^{nilp} \to \Gamma \otimes k$ is injective.*

PROOF. As every group is a projective limit of finitely generated groups, it is enough to prove the claim for finitely generated $\Gamma$. By the same reduction, one only needs to prove that every finitely generated torsion–free nilpotent group embeds in a $k$–unipotent group. This is done for example in an Appendix to [103]. □

The image of the homomorphism $\Gamma_0^{nilp} \to \Gamma \otimes k$ is, morally speaking, an integral form of $\Gamma \otimes k$, as the following lemma on the nilpotent steps of the tower shows.

LEMMA A.24. *If $\Gamma$ is a finitely presentable nilpotent group, then every element $g \in \Gamma \otimes \mathbb{Q}$ has the property that there exists an $n$ such that*

$$g^n \in \Gamma_0^{nilp} \subset \Gamma \otimes \mathbb{Q}.$$

PROOF. See the Appendix to [103]. □

The categories of pro–$k$–unipotent groups are partially ordered by the field extensions $K|k$, with the inclusion $U_k \subset U_K$ given by extension of scalars. The relation between the $k$–unipotent completions for different $k$ is similarly given:

LEMMA A.25. *If $\Gamma$ is finitely presentable, given a field extension $K|k$, there are $K$-Lie algebra isomorphisms*

$$(\mathcal{L}(\Gamma, k)) \underset{k}{\otimes} K \cong \mathcal{L}(\Gamma, K),$$

*which are natural in $\Gamma$.*

This is *not* a direct consequence of the universal property and categorical diagram chasing; it follows from properties of unipotent groups. For its proof, we refer to [71] or to [66][4].

Lemma A.25 implies that there are natural isomorphisms

$$\mathcal{L}(\Gamma, k) \cong \mathcal{L}(\Gamma, \mathbb{Q}) \underset{\mathbb{Q}}{\otimes} k.$$

Thus all the $k$–unipotent completions are obtained from the $\mathbb{Q}$–unipotent completion by extension of scalars.

By Remark A.21, $\Gamma \otimes k$ is a coarser invariant of $\Gamma$ than $\Gamma_0^{nilp}$. The comparison between the two completions is easier if one is given a homomorphism $f\colon \Gamma \to \Delta$ with good properties, such as those induced by proper holomorphic maps.

LEMMA A.26. *Let $f\colon \Gamma \to \Delta$ be a homomorphism between finitely presentable groups.*

*(i) If $Im(f) \subset \Delta$ has finite index, then $f \otimes k$ is an isomorphism if and only if $f_0^{nilp}$ is injective and has finite index image.*

*(ii) If $f$ is surjective, then $f \otimes k$ is an isomorphism if and only if $f_0^{nilp}$ is.*

PROOF. By the extension of scalars property of Lemma A.25 it suffices to prove the case of $k = \mathbb{Q}$.

(i) Let $f \otimes \mathbb{Q}$ be an isomorphism. As $\Gamma_0^{nilp} \hookrightarrow \Gamma \otimes \mathbb{Q}$ is injective, the map $f_0^{nilp}$ must also be injective. Moreover, the fact that $Im(f) \subset \Delta$ has finite index implies that the induced homomorphisms

$$(\Gamma/\Gamma_n)/_{Tor} \longrightarrow (\Delta/\Delta_n)/_{Tor}$$

---

[4]We owe this reference to R. Hain.

have images with finite index. Consequently, there is a tower of surjective homomorphisms from the finite quotient $\Delta/\mathrm{Im}(f)$ to the projective system of quotients

$$\cdots \longrightarrow ((\Delta/\Delta_n)/_{Tor})/\mathrm{Im}(f) \longrightarrow \cdots$$

These homomorphisms extend to a surjection from $\Delta/\mathrm{Im}(f)$ to its projective limit $\Delta_0^{nilp}/\mathrm{Im}(f_0^{nilp})$, which must therefore be finite.

Conversely, assume that $f_0^{nilp}$ is injective and almost surjective. This implies that all the finite steps of the projective system

$$(f_0^{nilp})_n : (\Gamma/\Gamma_n)/_{Tor} \longrightarrow (\Delta/\Delta_n)/_{Tor}$$

must also be injective and almost surjective homomorphisms. Consider now the projective system of maps

$$(f \otimes \mathbb{Q})_n : (\Gamma/\Gamma_n) \otimes \mathbb{Q} \longrightarrow (\Delta/\Delta_n) \otimes \mathbb{Q} \, .$$

By Lemma A.24 every element of $\ker(f \otimes \mathbb{Q})_n$ has a power in $\ker(f_0^{nilp})_n = \{1\}$. The groups $(\Gamma/\Gamma_n) \otimes \mathbb{Q}$ are torsion free, so the homomorphisms $(f \otimes \mathbb{Q})_n$ must be injective. Moreover, every element of $(\Delta/\Delta_n) \otimes \mathbb{Q}$ has a power in $(\Delta/\Delta_n)/_{Tor}$, thus also a possibly higher power in $\mathrm{Im}(f_0^{nilp})_n \subset \mathrm{Im}(f \otimes \mathbb{Q})_n$. As $(f \otimes \mathbb{Q})_n$ is a homomorphism of $\mathbb{Q}$-unipotent groups, this means that it is onto.

The completion $f \otimes \mathbb{Q}$ is the projective limit of the isomorphisms $(F \otimes \mathbb{Q})_n$, so it must also be an isomorphism.

(ii) can be proved analogously. □

Although the categorical inclusions might lead one to expect the opposite, the extension of scalars property of Lemma A.25 shows that field extensions $K|k$ coarsen the isomorphism type of $k$–unipotent completions. This is illustrated by the following example, which is Remark II.2.15 in [104]:

EXAMPLE A.27. For any field $k$, the *Heisenberg algebra* of dimension 3 with coefficients in $k$ is the Lie algebra

$$\mathfrak{h}_3(k) = \left\{ \begin{pmatrix} 0 & x & z \\ 0 & 0 & y \\ 0 & 0 & 0 \end{pmatrix} \in \mathfrak{gl}(3,k) \right\} .$$

The $\mathbb{Q}$–Lie algebras $\mathfrak{h}_1 = \mathfrak{h}_3(\mathbb{Q}[\sqrt{2}])$ and $\mathfrak{h}_2 = \mathfrak{h}_3(\mathbb{Q}) \times \mathfrak{h}_3(\mathbb{Q})$ are not isomorphic, but the $\mathbb{R}$–Lie algebras $\mathfrak{h}_1 \underset{\mathbb{Q}}{\otimes} \mathbb{R}$ and $\mathfrak{h}_2 \underset{\mathbb{Q}}{\otimes} \mathbb{R}$ are.

APPENDIX B

# A glossary of Hodge theory

The most fundamental tool in the study of the topology of Kähler manifolds is Hodge theory. In this appendix we collect a few basic facts of this theory, referring to the standard textbooks, e.g. [55], for proofs and further discussion.

Let $(X, \omega)$ be a compact Kähler manifold. We use the following notation:

CONVENTION B.1.  (i) $J$ denotes the complex structure $TX \to TX$ and its transpose $T^*X \to T^*X$,
(ii) the $d^c$ operator is defined as $d^c = J^{-1}dJ$,
(iii) the Laplacian of a map is denoted by $\Delta f = - * d_\nabla * df$, where $d_\nabla$ is the covariant derivative induced by the metrics on the domain and target manifolds.

The Kähler form $\omega$ is closed by definition, and thus defines a class in $H^2(X, \mathbb{C})$. As $\omega^n$ is a non–zero multiple of the volume form, we have

$$[\omega^k] \neq 0 \in H^{2k}(X, \mathbb{C}),$$

for all $k \leq n$, and thus:

FACT B.2. If $X$ is a compact Kähler manifold, its Betti numbers of even degree are non–zero.

DEFINITION B.3. A (pure) *Hodge structure* of weight $k$ consists of a free Abelian group $H_\mathbb{Z}$ of finite rank, together with a decomposition of its complexification:

(15) $$H_\mathbb{Z} \otimes_\mathbb{Z} \mathbb{C} = \oplus_{p+q=k} H^{p,q},$$

with the property that

$$H^{p,q} = \overline{H^{q,p}}$$

for all $p, q$.

Alternatively, a Hodge structure can be described as a decreasing filtration of $H_\mathbb{Z} \otimes_\mathbb{Z} \mathbb{C}$

(16) $$H_\mathbb{Z} \otimes_\mathbb{Z} \mathbb{C} = F^0 \supset F^1 \supset \ldots \supset F^k$$

such that

$$H_\mathbb{Z} \otimes_\mathbb{Z} \mathbb{C} = F^p \oplus \overline{F^{k-p+1}}.$$

The existence of the filtration is clearly equivalent to Definition B.3 because (15) leads to (16) by setting

$$F^p = H^{k,0} \oplus \ldots \oplus H^{p,k-p};$$

and conversely, (16) leads to (15) by setting

$$H^{p,q} = F^p \cap \overline{F^q}.$$

On a compact Kähler manifold, the equality

$$\Delta_d = 2\Delta_\partial = 2\Delta_{\bar{\partial}}$$

of Laplacians implies that the type decomposition of differential forms passes to cohomology and defines a Hodge structure of weight $k$ on $H^k(X,\mathbb{Z})/\text{torsion}$. In fact, the integral lattice does not play any role in this book, and we could just as well think of a Hodge structure as a type decomposition or filtration as above, defined on the complexification of a real vector space.

The Dolbeault isomorphism gives a natural interpretation of the Hodge decomposition of $H^k(X,\mathbb{C})$, by identifying the space of harmonic forms of type $(p,q)$ with the sheaf cohomology group $H^q_{\bar{\partial}}(\Omega^p_X)$. In particular, the spaces of holomorphic forms inject into the cohomology.

The existence of a pure Hodge structure of odd weight on a lattice implies that the rank of the lattice is even. Thus:

FACT B.4. *If $X$ is a compact Kähler manifold, its Betti numbers of odd degree are even:*

$$b_{2m+1}(X) \equiv 0 \pmod{2}.$$

On many occasions we use the following deeper result, which one can think of as a sharpening of Fact B.4:

THEOREM B.5 (Hard Lefschetz Theorem). *If $(X,\omega)$ is a compact Kähler manifold of complex dimension $n$, then multiplication by the cohomology class of $\omega$ defines an isomorphism*

$$\omega^k \colon H^{n-k}(X,\mathbb{C}) \longrightarrow H^{n+k}(X,\mathbb{C}),$$

*for every $k \in \{1,\ldots n\}$.*

COROLLARY B.6. *For all $k \leq n$, the following bilinear pairing is non–degenerate:*

$$H^k(X,\mathbb{C}) \times H^k(X,\mathbb{C}) \longrightarrow \mathbb{C}$$
$$(\alpha,\beta) \longmapsto \int_X \alpha \wedge \beta \wedge \omega^{n-k}.$$

PROOF. By Theorem B.5, multiplication by $\omega^{n-k}$ is an isomorphism if $k < n$. Then the claim follows from Poincaré duality. In the middle dimension $k = n$, the claim is just Poincaré duality. □

When $k = 1$, the Corollary is very useful in the study of Kähler groups. As we have seen in Example 1.20, the above pairing passes to the group cohomology of $\pi_1(X)$, which immediately leads to strong restrictions.

The operator $d^c = J^{-1}dJ$ is the real restriction of the operator $d^c = i(\bar{\partial} - \partial)$ on the Dolbeault complex $\mathcal{E}^*_X \otimes \mathbb{C}$. It satisfies the following identities:

(i) nilpotence: $(d^c)^2 = 0$,
(ii) commutativity with $d^c$: $dd^c = -d^c d = 2i\partial\bar{\partial}$,
(iii) Laplacian: $\Delta_{d^c} = \Delta_d = 2\Delta_\partial = 2\Delta_{\bar{\partial}}$.

The Laplacian identity implies the isomorphism of cohomology algebras $H^*(X,\mathbb{R}) = H^*_d(X) \cong H^*_{d^c}(X)$.

A crucial property of $d^c$ is:

LEMMA B.7 ($dd^c$ Lemma). *Let $X$ be a compact Kähler manifold, and let $\alpha$ be a differential form on $X$ such that $\alpha = d\gamma$. Then there exists a form $\beta$ such that $\alpha = dd^c\beta$.*

As $J$ and $dd^c$ commute, the operators $d$ and $d^c$ are actually interchangeable in the statement of the Lemma, this yields a "$d^c d$ Lemma".

# Bibliography

[1] S. Abhyankar, *Tame coverings and fundamental groups of algebraic varieties*, Amer. J. Math. **81** (1959), 46–94.

[2] J. Amorós, *On the Malcev completion of Kähler groups*, Comment. Math. Helv. (to appear).

[3] D. Arapura, *Higgs line bundles, Green–Lazarsfeld sets and maps of Kähler manifolds to curves*, Bull. of the Amer. Math. Soc. (New Series) **26** (1992) 310–314.

[4] D. Arapura, *Fundamental groups of smooth projective varieties*, (to appear).

[5] D. Arapura, P. Bressler and M. Ramachandran, *On the fundamental group of a compact Kähler manifold*, Duke Math. Jour. **68** (1992) 477–488.

[6] D. Arapura and M. Nori, *Solvable fundamental groups of algebraic varieties and Kähler manifolds*, (in preparation).

[7] M. F. Atiyah, N. J. Hitchin and I. M. Singer, *Self-duality in four-dimensional Riemannian Geometry*, Proc. Royal Soc. of London A **362** (1978) 425–461.

[8] W. Barth, C. Peters and A. Van de Ven, *Compact Complex Surfaces*, Springer Verlag 1984.

[9] A. Beauville, *Annulation du $H^1$ et systemes paracanoniques sur les surfaces*, J. Reine Angew. Math. **388** (1988) 149–157.

[10] A. Beauville, Appendix to [29].

[11] M. Bekka and A. Valette, *Group cohomology, harmonic functions, and the first $L^2$ Betti number*, Potential Analysis (to appear).

[12] S. Bloch, *Applications of the dilogarithm function in algebraic K-theory and algebraic geometry*, Proceedings of the International Symposium on Algebraic Geometry (Kyoto 1977), Kinokuniya Book Store, 1978.

[13] F. Bogomolov and L. Katzarkov, *Projective surfaces with interesting fundamental groups*, preprint 1995.

[14] A. Borel, *Compact Clifford-Klein forms of symmetric spaces*, Topology **2** (1963) 111–122.

[15] A. Borel, *Stable real cohomology of arithmetic groups*, Ann. Scient. Éc. Norm. Sup. **7** (1974) 235–272.

[16] A. Borel, *Cohomologie de sous-groupes discrets et représentations de groupes semi–simples*, Astérisque **32–33** (1976) 73–112.

[17] A. Borel, *Stable real cohomology of arithmetic groups II*, Manifolds and Lie Groups (Notre Dame, 1980), Progr. Math. **14**, Birkhäuser Verlag 1981.

[18] A. Borel and J.-P. Serre, *Corners and arithmetic groups*, Comment. Math. Helv. **48** (1973) 436–491.

[19] R. Bott, *On a theorem of Lefschetz*, Michigan Math. Journal **6** (1959) 211–216.

[20] J.-P. Bourguignon, *Eugenio Calabi and Kähler metrics*, Calabi Memorial Volume (to appear).

[21] A. Bousfield and V. Gugenheim, *On PL de Rham theory and rational homotopy type*, Mem. Amer. Math. Soc. **179**, Amer. Math. Soc., Providence, R.I. 1976.

[22] F. Campana, *Fundamental group and positivity of cotangent bundles of compact Kähler manifolds*, J. Alg. Geom. **4** (1995) 487–502.

[23] F. Campana, *Remarques sur les groupes de Kähler nilpotents*, Ann. Scient. Éc. Norm. Sup. **28** (1995) 307–316.

[24] J. A. Carlson and L. Hernández, *Harmonic maps from compact Kähler manifolds to exceptional hyperbolic spaces*, Jour. Geometric Analysis **1** (1991) 339–357.

[25] J. A. Carlson and D. Toledo, *Harmonic maps of Kähler manifolds to locally symmetric spaces*, Publ. Maths. IHES **69** (1989) 173–201.

[26] J. A. Carlson and D. Toledo, *Rigidity of harmonic maps of maximum rank*, Jour. Geometric Analysis **3** (1993) 99–140.

[27] J. A. Carlson and D. Toledo, *Quadratic presentations and nilpotent Kähler groups*, Jour. Geometric Analysis (to appear).

[28] J. A. Carlson and D. Toledo, *On fundamental groups of class VII surfaces*, Bull. London Maths. Soc. (to appear).

[29] F. Catanese, *Moduli and classification of irregular Kähler manifolds (and algebraic varieties) with Albanese general type fibrations*, Invent. Math. **104** (1991) 263–289.

[30] F. Catanese and J. Kollár, *Trento examples*, LNM **1515**, Springer Verlag 1992.

[31] J. Cheeger and J. Simons, *Differential characters and geometric invariants*, Geometry and Topology, Springer LNM **1167** (1985) 50–80.

[32] S. S. Chern and J. Simons, *Characteristic forms and geometric invariants*, Ann. of Math. **99** (1974) 48–69.

[33] J. Cheeger and M. Gromov, $L_2$-*cohomology and group cohomology*, Topology **25** (1986) 189–215.

[34] S. Chen, *Examples of n-step nilpotent 1-formal 1-minimal models*, C.R. Acad. Sci. Paris **321**, Série I (1995) 223–228.

[35] E. M. Chirka, *Complex Analytic Sets*, Mathematics and its Applications, Vol. **46**, Kluwer, Dordrecht 1989.

[36] K. Corlette, *Flat G-bundles with canonical metrics*, J. Differential Geometry **28** (1988) 361–382.

[37] K. Corlette, *Rigid repesentations of Kählerian fundamental groups*, J. Differential Geometry **33** (1991) 239–252.

[38] P. Deligne, *Poids dans la cohomologie des variétés algébriques*, Proc. Int. Cong. Math., Vancouver 1974 vol. **1** 79–85.

[39] P. Deligne, *Extensions centrales non résiduellement finies de groupes arithméthiques*, C. R. Acad. Sci. Paris, Série A-B, **287** (1978) 203–208.

[40] P. Deligne, *Un théorème de finitude pour la monodromie*, In: R. Howe (ed.) *Discrete Groups and Analysis*, Birkhäuser Verlag 1987.

[41] P. Deligne, P. Griffiths, J. W. Morgan and D. Sullivan, *Real homotopy theory of Kähler manifolds*, Invent. Math. **29** (1975) 245–274.

[42] J. Dodziuk, *de Rham–Hodge theory for $L^2$-cohomology of infinite coverings*, Topology **16** (1977) 157–165.

[43] S. K. Donaldson, *Twisted harmonic maps and the self-duality equations*, Proc. London Maths. Soc. **55** (1987) 127–131.

[44] J. Eells and L. Lemaire, *A report on harmonic maps*, Bull. London Maths. Soc. **10** (1978) 1–68.

[45] J. Eells and J. H. Sampson, *Harmonic maps of Riemannian manifolds*, Amer. Jour. Math. **86** (1964) 109–160.

[46] Y. Eliashberg, *Topological characterisation of Stein manifolds of dimension* > 2, Inter. Journ. of Math. **1** (1990) 29–46.

[47] D. B. A. Epstein, *Ends*, in *Topology of 3-manifolds and related topics*, Prentice Hall 1962, 110–117.

[48] R. Friedman and J. W. Morgan, *Smooth Four–Manifolds and Complex Surfaces*, Springer Verlag 1994.

[49] A. Fujiki, *Hyper–Kähler structure on the moduli space of flat bundles*, Prospects in Complex Geometry, Springer LNM **1468**, 1–83.

[50] W. M. Goldman, *The symplectic nature of fundamental groups of surfaces*, Adv. in Math. **54** (1984) 200–225.

[51] W. M. Goldman and J. J. Millson, *The deformation theory of representations of fundamental groups of compact Kähler manifolds*, Publ. Math. IHES **67** (1988) 43–96.

[52] R. E. Gompf, *A new construction of symplectic manifolds*, Ann. of Math. **142** (1995) 527–595.

[53] M. Goresky and R. MacPherson, *Stratified Morse Theory*, Springer Verlag 1988.

[54] M. Green and R. Lazarsfeld, *Deformation theory, generic vanishing theorems and some conjectures of Enriques, Catanese and Beauville*, Invent. Math. **90** (1987) 389–407.

[55] P. Griffiths and J. Harris, *Principles of Algebraic Geometry*, John Wiley & Sons, New York 1978.

[56] P. Griffiths and J. W. Morgan, *Rational Homotopy Theory and Differential Forms*, Progr. Math. **16**, Birkhäuser Verlag 1981.

[57] P. Griffiths and W. Schmid, *Locally homogeneous complex manifolds*, Acta Math. **123** (1969) 253–302.

[58] M. Gromov, *Sur le groupe fondamental d'une variété kählérienne*, C.R. Acad. Sci. Paris **308** Série I (1989) 67–70.

[59] M. Gromov, *Kähler hyperbolicity and $L^2$-Hodge theory*, J. Differential Geometry **33** (1991) 263–292.

[60] M. Gromov, *Foliated Plateau problem, part II: harmonic maps of foliations*, GAFA **1** (1991) 253–320.

[61] M. Gromov, *Asymptotic invariants of infinite groups*, in Geometric Group Theory, Vol. **2**, ed. G. A. Niblo and M. A. Roller, LMS Lecture Notes Series 182, Cambridge Univ. Press 1993.

[62] M. Gromov and R. Schoen, *Harmonic maps into singular spaces and p-adic superrigidity for lattices in groups of rank one*, Publ. Math. IHES (1992) 165–246.

[63] A. Guichardet, *Cohomologie des groupes topologiques et des algèbres de Lie*, CEDIC/Fernand Nathan, Paris 1980.

[64] R. Gunning, *Introduction to holomorphic functions of several variables*, Vol. 2, Wadsworth & Brooks/Cole 1990.

[65] R. Hain, *The de Rham homotopy theory of complex algebraic varieties*, K-theory **1** (1987) 271–324.

[66] R. Hain, *Completions of mapping class groups and the cycle $C - C^-$*, in C.-F. Bödigheimer, R. Hain (Eds.), Mapping Class Groups and Moduli Spaces of Riemann Surfaces, Contemporary Mathematics **150**, Amer. Math. Soc. 1993.

[67] R. Harvey and H. B. Lawson Jr., *An intrinsic characterisation of Kähler manifolds*, Invent. math. **74** (1983) 169–198.

[68] L. Hernández, *Kähler manifolds and 1/4 pinching*, Duke Math. Jour. **62** (1991) 601–611.

[69] G. Higman, *A finitely related group with an isomorphic factor group*, Jour. London Maths. Soc. **26** (1951) 59–61.

[70] G. Higman, *A finitely generated infinite simple group*, Jour. London Maths. Soc. **26** (1951), 61–64.

[71] P. Hilton, G. Mislin and J. Roitberg, *Localization of Nilpotent Groups and Spaces*, Math. Studies **15**, North Holland 1975.

[72] N. J. Hitchin, *The self-duality equations on a Riemann surface*, Proc. London Maths. Soc. **55** (1987) 59–126.

[73] N. J. Hitchin, *Stable bundles and integrable systems*, Duke Math. J. **54** (1987) 91–114.

[74] G. P. Hochschild and J.-P. Serre, *Cohomology of Lie algebras*, Ann. of Math. **57** (1953) 591–603.

[75] F. E. A. Johnson and E. G. Rees, *On the fundamental group of a complex algebraic manifold*, Bull. London Math. Soc. **19** (1987) 463–466.

[76] F. E. A. Johnson and E. G. Rees, *The fundamental groups of an algebraic variety*, in S. Jackowsky et al.(Eds.) Algebraic Topology Poznan 1989, LNM **1474**, Springer Verlag 1991.

[77] J. Jost, *Nonlinear Methods in Riemannian and Kählerian Geometry*, DMV Seminar **10**, Birkhäuser Verlag 1988.

[78] E. Kähler, *Über eine bemerkenswerte Hermitesche Metrik*, Abh. Math. Sem. Univ. Hamburg **9** (1933) 173–186.

[79] L. Katzarkov, *On the Shafarevich maps*, preprint 1995.

[80] J. Kollár, *Shafarevich maps and plurigenera of algebraic varieties*, Invent. Math. **113** (1993) 177–215.

[81] J. Kollár, *Shafarevich Maps and Automorphic Forms*, Princeton University Press 1995.

[82] N. Korevaar and R. Schoen, *Sobolev spaces and harmonic maps for metric space targets*, Comm. Analysis and Geom. **1** (1993), 561–660.

[83] J.-L. Koszul and B. Malgrange, *Sur certaines structures fibrées complexes*, Archiv Math. **9** (1958) 102–109.

[84] D. Kotschick, *Remarks on geometric structures on compact complex surfaces*, Topology **31** (1992) 317–321.

[85] D. Kotschick, *All fundamental groups are almost complex*, Bull. London Math. Soc. **24** (1992) 377–378.

[86] D. Kotschick, *Four-manifold invariants of finitely presentable groups*, in Topology, Geometry, and Field Theory, ed. K. Fukaya et. al., World Scientific 1994.

[87] F. Labourie, *Existence d'applications harmoniques tordues à valeurs dans les variétés à courbure négative*, Proc. Amer. Math. Soc. **111** (1991) 877–882.

[88] R. C. Lyndon and P. C. Schupp, *Combinatorial Group Theory*, Springer Verlag 1977.

[89] S. Mac Lane, *Categories for the Working Mathematician*, GTM 5, Springer Verlag 1971.

[90] G. A. Margulis, *Discrete Subgroups of Semisimple Lie Groups*, Springer Verlag 1991.

[91] N. Mihalache, *Every finitely presented group is the $\pi_1$ of some two dimensional Stein space*, Math. Annalen **298** (1994) 533–542.

[92] C. F. Miller, *On group-theoretic decison problems and their classification*, Annals of Math. Studies **68**, Princeton University Press 1971.

[93] J. J. Millson, *Real vector bundles with discrete structure group*, Topology **18** (1979) 83–89.

[94] J. W. Morgan, *The algebraic topology of smooth algebraic varieties*, Publ. Math. IHES **48** (1978) 137–204.

[95] I. Nakamura, *Towards classification of non-Kählerian complex surfaces*, Sugaku Expositions **2** (1989) 209–229.

[96] T. Napier and M. Ramachandran, *Structure theorems for complete Kähler manifolds and applications to Lefschetz type theorems*, GAFA **5** (1995) 809–851.

[97] T. Napier and M. Ramachandran, *Weakly 1-complete Kähler manifolds with large automorphism group*, preprint 1995.

[98] V. Navarro Aznar, *Sur la théorie de Hodge–Deligne*, Invent. Math. **90** (1987) 11–76.

[99] B. H. Neumann, *A two-generator group isomorphic to a proper factor group*, Jour. London Maths. Soc. **25** (1950) 247–248.

[100] P. Orlik, *Introduction to Arrangements*, CBMS Reg. Conference Series **72**, Amer. Math. Soc. 1989.

[101] Ch. Pittet, *Ends and isoperimetry*, preprint 1995.

[102] A. Polombo, *Nombres caractéristiques d'une variété riemannienne de dimension 4*, J. Differential Geometry **13** (1978) 145–162.

[103] D. Quillen, *Rational homotopy theory*, Ann. of Math. **90** (1969) 205–295.

[104] M. S. Raghunathan, *Discrete Subgroups of Lie Groups*, Ergebnisse der Math. **68**, Springer-Verlag 1972.

[105] M. S. Raghunathan, *Torsion in cocompact lattices in coverings of $Spin(2,n)$*, Math. Annalen **266** (1984), 403–419; Corrigendum, Math. Annalen **303** (1995) 575–578.

[106] A. Reznikov, *All regulators of flat bundles are torsion*, Ann. of Math. **141** (1995) 373–386.

[107] G. de Rham, *Variétés différentiables*, Act. Scient. et Ind. **1222**, Hermann 1960.

[108] J. H. Sampson, *Some properties and applications of harmonic mappings*, Ann. Scient. Éc. Norm. Sup. **11** (1978) 211–228.

[109] J. H. Sampson, *Applications of harmonic maps to Kähler geometry*, in Complex Differential Geometry and Nonlinear Differential Equations, ed. Y.-T. Siu, Contemp. Math. **49** 1986.

[110] H. Seifert und W. Threlfall, *Lehrbuch der Topologie*, Teubner Verlag, Leipzig 1934.

[111] Séminaire Arthur Besse, *Géométrie Riemannienne en dimension 4*, CEDIC/Fernand Nathan, Paris 1981.

[112] J.-P. Serre, *Sur la topologie des variétés algébriques en caractéristique p*, Symp. Int. Top. Alg., Mexico (1958) 24–53; reprinted in Collected Papers, Vol. 1, Springer Verlag 1986.

[113] J.-P. Serre, *Trees*, Springer Verlag 1980.

[114] I. R. Shafarevich, *Basic Algebraic Geometry*, Springer Verlag 1977.

[115] C. T. Simpson, *Constructing variations of Hodge structure using Yang–Mills theory and applications to uniformization*, Journal of the AMS **1** (1988) 867–918.

[116] C. T. Simpson, *Higgs bundles and local systems*, Publ. Math. IHES **75** (1992) 5–95.

[117] C. T. Simpson, *A Lefschetz theorem for $\pi_0$ of the integral leaves of a holomorphic one-form*, Compos. Math. **87** (1993) 99–113.

[118] C. T. Simpson, *Moduli of representations of the fundamental group of a smooth projective variety I*, Publ. Math. IHES **79** (1994) 47–129.

[119] C. T. Simpson, *Moduli of representations of the fundamental group of a smooth projective variety II*, Publ. Math. IHES **80** (1994) 5–79.

[120] Y.-T. Siu, *The complex-analyticity of harmonic maps and the strong rigidity of compact Kähler manifolds*, Annals of Math. **112** (1980) 73–111.

[121] Y.-T. Siu, *Complex analyticity of harmonic maps, vanishing and Lefschetz theorems*, J. Differential Geometry **17** (1982) 55–138.

[122] Y.-T. Siu, *Strong rigidity for Kähler manifolds and the construction of bounded holomorphic functions*, In: R. Howe (ed.) *Discrete Groups and Analysis*, Birkhäuser Verlag 1987.

[123] P. Soardi and W. Woess, *Amenability, unimodularity, and the spectral radius of random walks on infinite graphs*, Math. Zeit. **205** (1990) 471–486.

[124] A. Sommese and A. Van de Ven, *Homotopy groups of pullbacks of varieties*, Nagoya Math. Jour. **102** (1986) 79–90.

[125] J. Stallings, *Homology and central series of groups*, J. Algebra **2** (1965) 170–181.

[126] J. Stallings, *Group theory and three-dimensional manifolds*, Yale Math. Monographs **4**, Yale Univ. Press 1971.

[127] K. Stein, *Un théorème sur le prolongement des ensembles analytiques*, Séminaire Cartan, E.N.S., Exposé 13, 1954.

[128] K. Stein, *Analytische Zerlegung komplexer Räume*, Math. Annalen **132** (1956), 63–93.

[129] K. Stein, *On factorization of holomorphic mappings*, Proc. of the Conference on Complex Analysis, Minneapolis 1964, Springer Verlag 1965.

[130] D. Sullivan, *Infinitesimal computations in topology*, Publ. Math. IHES **47** (1977) 269–332.

[131] C. H. Taubes, *The existence of anti-self-dual conformal structures*, J. Differential Geometry **36** (1992) 163–253.

[132] D. Toledo, *Examples of fundamental groups of compact Kähler manifolds*, Bull. London Maths. Soc. **22** (1990) 339–343.

[133] D. Toledo, *Projective varieties with non-residually finite fundamental group*, Publ. Maths. IHES **77** (1993) 103–119.

[134] K. Ueno, *Classification Theory of Algebraic Varieties and Compact Complex Surfaces*, Lecture Notes in Math. **439**, Springer-Verlag 1975.

[135] P. D. Watson, *On the limits of sequences of sets*, Oxford Quart. J. Maths. **4** (1953) 1–3.

[136] C. Weber (ed.), *Nœuds, tresses et singularités*, Monographie No 31 de L'Enseignement Mathématique, Genéve 1983.

[137] O. Zariski, *Algebraic Surfaces* (Second Supplemented Edition), Springer Verlag 1971.

[138] K. Zuo, *Factorization of nonrigid Zariski dense representations of $\pi_1$ of projective manifolds*, Invent. Math. **118** (1994) 37–46.

# Index

1–formal space, 34, 36–37
1–minimal model, 29–32, 37, 41

Albanese
  dimension, 23
  image, 23, 40–43
  map, 3, 22–25, 40–43
  variety, 22–23, 41

Betti moduli space, 92–94
Bochner formula, see Siu–Sampson theorem
bounded geometry, 48, 50, 53–63

Castelnuovo–de Franchis, 2–3, 24–25, 28, 43, 61, 63, 90
Cheeger–Chern–Simons classes, 106
commutative differential graded algebra (CDGA), 30–32, 35–37
  dual Lie algebra of, 32
  quasi–isomorphism of, 30
  weak equivalence of, 30, 32
compact complex surface, 4–5, 10–16, 25, 28

$dd^c$ lemma, 33, 131
de Rham fundamental group, 29–46, 123
  2–step nilpotent, 38

Eilenberg–Mac Lane space, 2–3, 8, 13, 21, 23, 33–34, 82–83
ends of groups, 8–9, 12–13, 47–51, 60–61

factorisation theorem, 26, 82–84, 89–90, 112
formality, 29, 32–34, 104–105
  1–formal space, 34, 36–37

group
  1–relator, 39
  Abelian, 9, 116, 123–125
  (pure) braid, 8, 61
  fibered, 27, 43, 86
  finite, 6
  free, 7, 9, 11, 13, 16, 18, 27, 38, 49, 123
  free product, 8, 11–14, 47, 49, 109–110
  Heisenberg, 8, 39–40, 49, 114
  Higman, 7, 14, 85–87, 110
  lattice, see lattice
  nilpotent, 9, 114
  non–fibered, 27, 43–45
  non–Hopfian, 110
  non–linear, 85–87, 110, 115–119
  non–residually finite, 7, 14, 109–110, 115–119
group algebra, 124–125
group completion, 121–128
  $k$–unipotent, see pro–unipotent
  $\mathcal{P}$–completion, 121
  nilpotent, 122, 124–128
  pro–finite, 7, 118, 122
  pro–unipotent, 9, 29–46, 123–128
  torsion–free nilpotent, 9, 43, 124–128

Hard Lefschetz Theorem, 8, 38, 130
harmonic flat bundle, 91–107
  deformation theory of, 104–105
harmonic map, 9–10, 26, 65–90
  existence of, 66, 68, 89
  equivariant, 67–69, 82, 92
  factorisation of, 26, 82, 89–90
  uniqueness of, 66
Hermitian sectional curvature, 71–76
Hermitian symmetric space, 78–81, 85
Higgs bundle, 91–105
Higgs–Hermitian–Yang–Mills (HHYM) equation, 97–103

Hodge signature theorem, 73
Hodge structure, 81, 91, 129–130
   mixed, 37, 45–46
   variation of, 94–95, 107
holomorphic foliation, 24, 54–63, 83–84, 90
hyperbolic manifold, 8, 13–14, 77–78, 82–85
hyperkähler manifold, 100–103

isoperimetric inequality, 50–51

$L^2$-cohomology, 8, 47–63
   exact, 48
   reduced, 48
lattice, 7, 81–83, 85, 87, 95–96, 110–111, 115–119
Lefschetz hyperplane theorem, 5–6, 11, 111–113, 118
Lie algebra, 76–81
   deformation of, 104
   differential graded (DGLA), 104–105
   dual of CDGA, 32
   lower central series of, 34
   quadratically presented, 34, 36

Malcev algebra, 29, 34–40, 125
   $n$–step nilpotent, 38, 126
   quadratic presentation of, 34–38
Massey product, 8, 38–40

normal variety, 9–10

open problem, 4, 7, 9, 85, 116

period domain, 81, 95
pluriharmonic map, 10, 71–90
   equivariant, 72, 93

quasi-isomorphism
   of CDGAs, 30
   of DGLAs, 104–105

Siu–Beauville theorem, 2–3, 8, 25–28, 43
Siu rigidity theorem, 80
Siu–Sampson theorem, 71–76, 89, 92–93, 96, 107
symplectic
   form, 3, 5, 18
   structure, 3, 5, 16, 18
   manifold, 5, 15–20, 81
   sum 17–18

**ISBN 0-8218-0498-7**